# Introduction to Aviation Safety

## 1st Edition

Brandon Wild, Gary Ullrich,
and Ryan Guthridge

**Kendall Hunt**
publishing company

**Kendall Hunt**
publishing company

www.kendallhunt.com
*Send all inquiries to:*
4050 Westmark Drive
Dubuque, IA 52004-1840

# Contents

| | | |
|---|---|---|
| | *About the Authors* | *v* |
| | *Acknowledgments* | *vii* |
| **1** | The Philosophy of Safety | 1 |
| **2** | Pilot Decision Making | 15 |
| **3** | Hazardous Attitudes | 35 |
| **4** | Managing Fatigue | 41 |
| **5** | Mid Air Collision Avoidance | 47 |
| **6** | Bird Strike Mitigation | 61 |
| **7** | Hazardous Weather | 67 |
| **8** | The Human Factors Analysis and Classification System (HFACS) and the "Swiss Cheese Model of Accident Causation" | 89 |
| **9** | Accident Investigation Theory | 103 |
| **10** | Aircraft Accident Investigation | 111 |
| **11** | Aviation Safety Program Management and Safety Management System (SMS) | 123 |
| **12** | Flight Safety Programs | 137 |
| **13** | Anti-Drug Programs in Aviation | 145 |
| **14** | General Aviation Accidents | 155 |
| **15** | Air Traffic Control (ATC) Safety | 169 |
| | *Key Terms* | *193* |
| | *Appendix A: Production versus Protection: Fierce Competition and Flawed Culture at Boeing* | *195* |

## BRANDON WILD, PH.D.

Brandon Wild is an Associate Professor in the Aviation Department at the University of North Dakota (UND). Prior to joining UND, Brandon held Flight Operational Quality Assurance (FOQA) analyst positions with both United Airlines and UPS, was an aviation underwriter at USAIG, and FOQA manager in the Flight Safety Department at Northwest Airlines and Delta Air Lines. He was also previously an adjunct instructor at Embry-Riddle Aeronautical University (ERAU).

Brandon holds a Bachelor of Science in Aerospace Studies and a Master of Business Administration in Aviation, both from Embry-Riddle Aeronautical University, and a Ph.D. in Educational Foundations and Research from the University of North Dakota.

## GARY ULLRICH, M.S.

Gary Ullrich is an Associate Professor in the Aviation Department at the University of North Dakota (UND). Prior to joining UND, Gary held positions as a pilot instructor and quality assurance specialist/inspector with FlightSafety, instructor pilot and check airman with the United States Air Force, aviation and ground accident investigator, and the Chief of Safety where he managed aviation, ground, ergonomics, and nuclear safety programs for an Air Force Wing of over 12,000 personnel. He holds FAA type ratings in the Boeing 707 and 720 along with multiple OSHA outreach training credentials.

Gary received a Bachelor of Science in Civil Engineering from North Dakota State University, Master of Science in Aeronautical Science from Embry-Riddle Aeronautical University, and a Safety and Health Specialist Certificate from the University of California, San Diego.

## RYAN GUTHRIDGE, M.B.A.

Ryan Guthridge is an Assistant Professor in the Aviation Department at the University of North Dakota (UND). Prior to joining UND, Ryan held positions as a Pilot-in-Command for Weather Modification, Inc., Technical Pilot for GE Aviation, Flight Data Analyst for Austin Digital, Inc., CEO for a UAS start-up, and Chief Ground Instructor for the University of Nebraska at Omaha. Through his background in data analytics, Ryan has analyzed flight data to identify events hazardous to flight safety, quantified the impact of procedures-based changes on flight efficiency, and assessed the performance of pilots in skills-based flight training maneuvers. Ryan is an FAA-certified Instrument Flight Procedure Validation pilot and holds FAA type ratings in the Airbus A320 and Boeing B737.

Ryan received a Bachelor of Science in Aeronautics from the University of North Dakota, Master of Business Administration from the University of Texas at Dallas, and Professional Certificates in Business Intelligence, Data Mining, and Learning Analytics.

# Acknowledgments

The authors would like to thank their amazing wives, Elizabeth Wild, Mary Ullrich, and Andrea Guthridge, for their support during the development of this book. Additionally, the authors would like to thank Terra Jorgensen for her contribution of Chapter 15 on Air Traffic Control Safety along with Michael Ullrich for his graphics support.

# The Philosophy of Safety

## Is Safety a Core business function?

In successful aviation organizations, the management of safety is a core business function – as is financial management. We often hear aviation professionals tell us that nothing is more important than safety. Can safety really be the number one objective? Probably not. Successful aviation organizations establish an effective safety management that has a realistic balance between safety and production goals. The finite limits of personnel, time, resources, financing, and operational performance must be accepted in any industry. If properly implemented, safety management maximizes both safety and the operational effectiveness of an organization. Safety must co-exist with our production objectives. There is no aviation organization that has been created to deliver only safety.

A misperception has been pervasive in aviation regarding where safety fits, in terms of priority, within the organization. This misperception has evolved into a universally accepted stereotype: in aviation, safety is the first priority. While socially, ethically, and morally impeccable; the stereotype and the perspective that it conveys does not hold ground when considered from the perspective that the management of safety is an organizational process.

All aviation organizations, regardless of their nature, have a business component with production goals (as shown in Figure 1.1). An Air Traffic Control Facility may have a production goal of 100 aircraft operations per hour. An airport may have a production goal of 100 operations per hour,

**Figure 1.1** Cost vs. Benefit

*Source:* FAA.gov

using parallel runways, under IFR conditions. A military organization may have a production goal of bombs-on-target anywhere in the world in 24 hours or less. Thus, all aviation organizations can be considered business organizations with production goals. A simple question is then relevant to shed light on the truthfulness, or lack thereof, of the safety stereotype: what is the fundamental objective of a business organization? The answer to this question is obvious: to deliver the service for which the organization was created in the first place, to achieve production objectives and eventually deliver dividends to stakeholders.

## Cost vs. Benefit Considerations

Operating a profitable, yet safe airline or service provider requires a constant balancing act between the need to fulfil production goals (such as departures that are on time) versus safety goals (such as taking extra time to ensure that a door is properly secured). The aviation workplace is filled with potentially unsafe conditions which will not all be eliminated; yet, operations must continue.

Some operations adopt a goal of *"zero accidents"* and state *"safety is their number one priority"*. The reality is that operators (and other commercial aviation organizations) need to generate a profit to survive. Profit or loss is the immediate indicator of the company's success in meeting its production goals. However, safety is a prerequisite for a sustainable aviation business, as a company tempted to cut corners will eventually realize. For most companies, safety can best be measured by the absence of accidental losses. Companies may realize they have a safety problem following a major accident or loss, in part because it will impact on the profit/loss statement. However, a company may operate for years with many potentially unsafe conditions without adverse consequence. Without effective safety management to identify and correct these unsafe conditions, the company may assume that it is meeting its safety objectives, as evidenced by the *"absence* of losses". In reality, it has been lucky.

Safety and profit are not mutually exclusive. Indeed, quality organizations realize that expenditures on the correction of unsafe conditions are an investment towards long-term profitability. Losses cost money. As money is spent on risk reduction measures, costly losses are reduced (as shown in Figure 1.2). However, by spending more and more money on risk reduction, the gains made through reduced losses may not be in proportion to the expenditures. Companies must balance the costs of losses and expenditures on risk reduction measures. Some level of loss may be acceptable from a straight profit and loss point of view; however, few organizations can survive the economic consequences of a major accident. Hence, there is a strong economic case for an effective Safety Management System (SMS) to manage the risks.

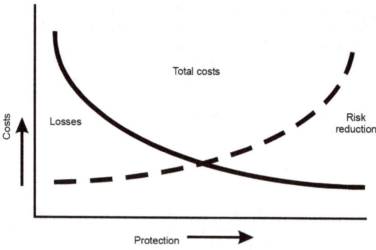

**Figure 1.2** Total Costs vs Protection
*Source:* Authors

## Costs of accidents

There are two basic types of costs associated with an accident or a serious incident: direct and indirect costs.

### Direct costs

These are the obvious costs, which are easy to determine. They mostly relate to physical damage and include rectifying, replacing or compensating for injuries, aircraft equipment and property damage. The high costs of an accident can be reduced by insurance coverage. (Some large organizations effectively self-insure by putting funds aside to cover their risks.)

### Indirect costs

While insurance may cover specified accident costs, there are many uninsured costs. An understanding of these uninsured costs (or indirect costs) is fundamental to understanding the economics of safety.

Indirect costs include all those items that are not directly covered by insurance and usually total much more than the direct costs resulting from an accident. Such costs are sometimes not obvious and are often delayed. Some examples of uninsured costs that may accrue from an accident include:

- *Loss of business and damage to the reputation of the organization.* Many organizations will not allow their personnel to fly with an operator with a questionable safety record.

- *Loss of use of equipment.* This equates to lost revenue. Replacement equipment may have to be purchased or leased. Companies operating a one-of-a-kind aircraft may find that their spares inventory and the people specially trained for such an aircraft become surplus.
- *Loss of staff productivity.* If people are injured in an accident and are unable to work, many states require that they continue to be paid. Also, these people will need to be replaced at least for the short term, incurring the costs of wages, overtime (and possibly training), as well as imposing an increased workload on the experienced workers.
- *Investigation and clean-up.* These are often uninsured costs. Operators may incur costs from the investigation including the costs of their staff involvement in the investigation, as well as the costs of tests and analyses, wreckage recovery, and restoring the accident site.
- *Insurance deductibles.* The policyholder's obligation to cover the first portion of the cost of any accident must be paid. A claim will also put a company into a higher risk category for insurance purposes and therefore may result in increased premiums. (Conversely, the implementation of a comprehensive SMS could help a company to negotiate a lower premium.)
- *Legal action and damage claims.* Legal costs can accrue rapidly. While it is possible to insure for public liability and damages, it is virtually impossible to cover the cost of time lost handling legal action and damage claims.
- *Fines and citations.* Government authorities may impose fines and citations, including possibly shutting down unsafe operations.

## Costs of Incidents

Serious aviation incidents, which result in minor damage or injuries, can also incur many of these indirect or uninsured costs. Typical cost factors arising from such incidents can include:

- Flight delays and cancellations;
- Alternate passenger transportation, accommodation, complaints, etc.;
- Crew change and positioning;
- Loss of revenue and reputation;
- Aircraft recovery, repair and test flight; and
- Incident investigation.

## Costs of safety

The costs of safety are even more difficult to quantify than the full costs of accidents. This is partly because of the difficulty in assessing the value of accidents that have been prevented. Nevertheless, some operators have attempted to quantify the costs and benefits of introducing an Safety Management Systems (SMS). They have found the cost savings to be substantial. Performing a cost-benefit analysis is complicated; however, it is an exercise that should be undertaken, as senior management is not inclined to spend money if there is no quantifiable benefit. One way of addressing this issue is to separate the costs of managing safety from the costs of correcting safety deficiencies, by charging the safety management costs to the safety department, and the safety deficiency costs to the line management most responsible. This exercise requires senior management's involvement in considering the costs and benefits of managing safety.

In successful aviation organizations, safety management is a core business function – as is financial management. Effective safety management requires a realistic balance between safety and production goals. Thus, a coordinated approach in which the organization's goals and resources are analyzed helps to ensure that decisions concerning safety are realistic and complementary to the operational needs of the organization. The finite limits of financing and operational performance must be accepted in any industry. Defining acceptable and unacceptable risks is therefore important for cost-effective safety management. If properly implemented, safety management measures not only increase safety but also improve the operational effectiveness of an organization.

## What is the value of a human life?

The United States government has conducted several studies on the treatment of the economic value of a statistical life (VSL). The FAA is organized under the Department of Transportation (DOT), so we will concentrate our emphasis on the DOT's value of a statistical life.

The DOT recognized VSL has a major effect on policy for the FAA. Due to federal law established by the United States Congress, the FAA is required to promote safety and air commerce. Ultimately this requires a cost/benefit analysis prior to making a new regulation. It is good to note that the National Transportation Safety Board (NTSB) does not conduct cost/benefit ratio analysis prior to issuing recommendations to the FAA.

The United States Department of Transportation (DOT) guidance on valuing reduction of fatalities and injuries by regulations or investments has been published periodically since 1993.

Empirical studies published in recent years indicate an average value of a statistical life (VSL) of $9.1 million in current U.S. dollars for analyses. Although the average value of a human life is $9.1 million, the DOT has established a variability of +/-$3.8 million. This ultimately establishes a range between $5.2 million up to $12.9 million.

## Value of Preventing Injuries

An accident which results in a loss in quality of life, including both pain and suffering and reduced income, should also be estimated. The dollar value for being injured will be less than the rate for the loss of a life. The fractions shown in Table 1.1 should be multiplied by the current VSL to obtain the values of preventing injuries of the types affected by the government action being analyzed.

For example, if the analyst were seeking to estimate the value of a "serious" injury (AIS 3), he or she would multiply the Fraction of VSL for a serious injury (0.105) by the VSL ($9.1 million) to calculate the value of the serious injury ($955,000). Values for injuries in the future would be calculated by multiplying these Fractions of VSL by the future values of VSL.

## Recognizing Uncertainty

Multiple studies have been conducted to determine the value of a human life. Some studies suggest a reasonable range of values for VSL between $4 million and $12.9 million. Additionally, different organizations within the U.S. government use different values. As an example, the Environmental Protection Agency uses a different value for VSL.

Because the relative costs and benefits of different provisions of a rule can vary greatly, it is important to disaggregate the provisions of a rule, displaying the expected costs and benefits of each provision, together with estimates of costs and benefits of reasonable alternatives to each provision.

## Production vs. Protection

In most organizations, safety is the number one priority, correct? That is what the public expects, and that is what most (if not all) organizations would like you to think. This is especially true in aviation. What aviation organization is not going to tell you that safety is their number one priority? If there is one, they are probably not going to be in business very long. Nobody is going to want to do business with an organization that admits that safety is not their number one priority, especially if the customer is going to be a passenger and has to actually fly on them!

**Table 1.1** Relative Disutility Factors by Accident Injury Severity Level (AIS)

| AIS Level | Severity | Fraction of VSL |
|-----------|----------|-----------------|
| AIS 1 | Minor | 0.003 |
| AIS 2 | Moderate | 0.047 |
| AIS 3 | Serious | 0.105 |

*Source:* Authors

However, saying that safety is your top priority and actually making it your top priority are two, totally separate items. Is safety really the number one priority for an organization? We would like to think that this is true, but in reality, it probably is not going to work out that way in the end. So, why not? To really answer this question we have to look at what the fundamental objective of any company or organization actually is. The fundamental objective of any organization is to meet their production goals or objectives. If they are a for profit corporation this could be taken one step further to say that that objective is to make a profit.

In either of those cases, is safety a competing priority? This can depend on the company. A quick definition of safety is: "the absence of risk." If an aviation organization is going to be completely absent from risk, they must stop flying their aircraft, because flying definitely involves risk. They are probably going to have to go one step further and park the airplanes in a remote area of the airport with a fence around them to prevent people from gaining access to them and having the possibility of injuring themselves. Okay, the last part may be a bit much, but you get the idea. Of course, the problem with this scenario is that it is unrealistic. If we park all our airplanes, we are not going to meet any of our goals or objectives, including making money. So, what can we do about this? One thing we are going to have to do is accept that there will be some risk in our operation. We are going to have to find a way to manage that risk, thereby letting us work on meeting our production objectives.

It seems that safety may not be able to be our top priority. But, can it be complimentary? There are plenty of airlines flying today, and the general public does not think twice about flying from point A to point B on them. But, airlines do not use safety in their marketing efforts. Why is it that passengers still fly on these airlines? It is because safety is assumed. Passengers expect that they will be safe on any given airline that they fly on. This is probably a realistic expectation, at least in areas of the world that have very robust oversight into airline safety.

In a company where safety is taken seriously it is treated as a core business function, much like accounting or marketing. Proper safety management will be able to properly manage the risk associated with operating aircraft or whatever the organization is doing. We need to find the "sweet" spot where the right balance is achieved between production and protection. See Figure 1.3 for a visual depiction of this safety zone.

If a corporation is to achieve the organizational objectives, but also be a safe organization, it will have to commit some of its resources into safety management. We will explore safety management systems

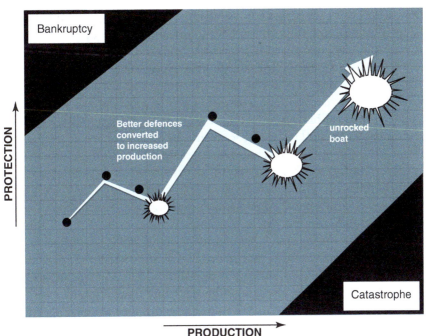

**Figure 1.3** The Safety Zone

(SMS) in a later chapter. However, just because an organization commits financial resources to safety, does not mean that the desired level of safety is achieved. The members of the organization from top to bottom have to believe in the safety objective. How is this achieved? Through a strong safety culture, which will be explored in the next section of Chapter 1.

## Safety Culture

In the United States, all airlines have to be certificated under Federal Aviation Regulation (FAR) Part 121 or Part 135. These regulations, enforced by the Federal Aviation Administration, outline the requirements that an airline must have in order to be granted an operating certificate. So, if that is the case, all airlines must meet the same requirements as any other. So, why do some airlines seem to have more safety issues than others? One of the reasons is that these airlines do not have a strong safety culture. What makes a strong safety culture?

One perspective is the belief that safety should be a core business function, as described in the earlier section on production vs. protection. But, this perspective cannot just be a perception of the safety department. This belief must be a core value throughout an organization, from top to bottom. A strong safety culture sends a message from the President down to the lowest level of employees that safety is an intrinsic value in the organization. If an employee does not see the levels of management talking about safety values, then the perception becomes one of not believing it is a core value in the company. This can be bad business in an organization like a bank or retail establishment. However, for an airline this could have deadly consequences. Think of the pilots, mechanics, ramp agents, or flight attendants. If they do not see (or believe) that management values or promotes safety, the belief may become one of apathy towards safety.

One way airlines and other aviation organizations have come to combat this attitude towards safety is through a system called "Just Culture." We know that mistakes will happen in an organization. A system of Just Culture promotes the idea that management will support employees for reporting these mistakes, instead of punishing them. Management teams who embrace this idea encourage the reporting of safety issues through some sort of reporting system. We will explore this type of reporting system in a later chapter.

In a system such as we have described above, employees are going to be much more willing to believe that management does value safety and that it is a core business function. Figure 1.4 shows a how Just Culture fits into an organization's safety culture.

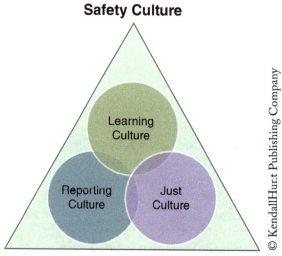

**Figure 1.4** Just Culture is a Part of Safety Culture

So, how do we define a safety culture in an organization? There have to be certain elements in an organization's culture to call it a safety culture. These elements are:

- An informed culture: people understand the hazards and risks involved in their own operations and all employees work continuously to identify and overcome threats to safety;
- A just culture: errors are understood but willful violations aren't tolerated; workers know and agree on what's acceptable and unacceptable;
- A reporting culture: workers are encouraged to voice safety concerns and when they do so, those concerns are analyzed and appropriate action is taken; and
- A learning culture: workers are encouraged to develop and apply their own skills and knowledge to enhance workplace safety; management updates workers on safety issues; safety reports are given to workers so that everyone learns the lessons.

Table 1.2 shows how an organization can figure out where there safety culture actually is.

Safety is somewhat of an enigma to a lot of aviation organizations. It is well known that it is a necessity, but how to actually have that safe operation becomes a goal that is economic in nature, and seems to somehow get pushed to the side for other, more highly (perceived) important organizational goals.

**Table 1.2** Safety Culture Characteristics

| Safety Culture: →  Characteristics | Poor | Bureaucratic | Positive |
|---|---|---|---|
| Hazard information is: | Suppressed | Ignored | Actively sought |
| Safety messengers are: | Discouraged or punished | Tolerated | Trained and encouraged |
| Responsibility for safety is: | Avoided | Fragmented | Shared |
| Dissemination of safety information is: | Discouraged | Allowed but discouraged | Rewarded |
| Failures lead to: | Cover-ups | Local fixes | Inquiries and systemic reform |
| New ideas are: | Crushed | Considered as new problems (not opportunities) | Welcomed |

*Source:* Authors

As we have seen in Chapter 1, in order to manage risk and obtain a culture of safety throughout the organization, it is important to think of safety along the same lines as the finance or human resources department. As we will explore in later chapters, there are many resources and ways for companies and organizations to increase their safety awareness.

## Chapter Questions

1. How important is a customer's perception of safety to an airline?
2. Why is it important for safety culture to begin at the top of the organization structure?
3. Do you think that a company could put too many resources into safety? What could be the result of committing too many resources to safety?
4. Which of the following statements concerning cost/benefit ratio analysis **is TRUE?**
   A) The "benefit" portion of cost/benefit ratio analysis includes loss of revenue, installation costs, and maintenance costs.
   B) The statistical value of a human life is determined by a congressional committee. This exact same statistical value of a human life is then used for all government agencies. (As an example: The Department of Transportation (DOT), and the Environmental Protection Agency (EPA) use the same value.
   C) The Federal Aviation Administration (FAA) is required to conduct a cost/benefit analysis prior to making any new regulations.
5. The term "Tombstone Technology refers to."
6. According to the Department of Transportation (DOT), what is the current statistical value of a human life?
7. True/False. ALL United States Government agencies forced to use the same statistical value of a human life. As an example, the Department of Transportation (DOT) and the Environmental Protection Agency (EPA) both use $8.0 million as the value of a human life.
8. True/False. Safety is widely used as an advertising and marketing tool with most every air carrier.
9. True/False. The FAA is allowed (because of laws passed by congress) to routinely establish new regulations that would result in the bankruptcy of most aviation companies.

10. Which of the following are the charters given to the FAA by the U.S. Congress:
    A) Promote a zero mishap rate
    B) Implement a "Zero Delay Program" within the Air Traffic Control System
    C) Promote air commerce and air safety
    D) Provide government oversight of the "Lost Baggage Program".

# Pilot Decision Making

---

**LEARNING OBJECTIVES**

1. Understand the FAA's definition of aeronautical decision making.
2. Understand how a brain continues to develop until age 25.
3. Understand risk management.
4. Define risk.
5. Know and understand the PAVE model for decision making.
6. Understand how an understanding of hazardous attitudes can help you make better decision.

---

Aeronautical decision-making (ADM) is decision-making in a unique environment – aviation. It is a systematic approach to the mental process used by pilots to consistently determine the best course of action in response to a given set of circumstances. It is what a pilot intends to do based on the latest information he or she has. The importance of learning and understanding effective ADM skills cannot be overemphasized. While progress is continually being made in the advancement of pilot training methods, aircraft equipment, aircraftsystems, and pilot services, accidents still occur. Despite the changes in technology to improve flight safety, one factor remains the same: the human factor which leads to errors. It is estimated that approximately 80 percent of all aviation accidents are related to human factors and the vast majority of these accidents occur during landing (24.1 percent) and takeoff (23.4 percent) (Figure 2.1). ADM is a systematic approach to risk assessment and stress management. To understand ADM is to also understand how personal attitudes can influence decision-making and how those attitudes can be modified to enhance safety in the flight deck. It is

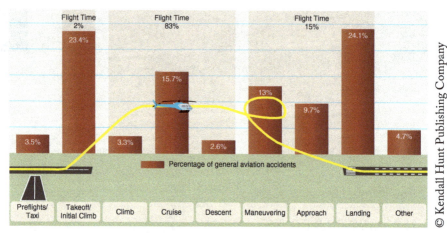

**Figure 2.1** Percentage of Aviation Accidents Related to Phases of Flight

important to understand the factors that cause humans to make decisions and how the decision-making process not only works, but can be improved. This chapter focuses on helping the pilot improve his or her ADM skills with the goal of mitigating the risk factors associated with flight. Advisory Circular (AC) 60-22, Aeronautical Decision-Making, provides background references, definitions, and other pertinent information about ADM training in the general aviation (GA) environment.

## Why do good people have accidents?

Whenever there's an accident, whether the result is a fatality, or a broken plate, or anything in between, someone is sure to ask, "How did it happen?"

The answer should always be the same: "It didn't happen, it was caused." And it is almost always possible to trace it back to somebody – or several somebodies – who failed to do their job somewhere along the line. Either they did something they shouldn't have done, or they failed to do something they should have done.

Let's suppose that you fall on the stairs at home and break a leg. That accident didn't "just happen." There was no evil spirit putting the hex on you or lurking in the shadows to trip you. No, there was at least one quite tangible cause.

The odds are that the fall was your own fault – that some act of yours (or failure to act) was to blame. Maybe you were in a hurry and took the stairs faster than usual – faster than was safe. Maybe you were carrying an awkward load that put you off balance and kept you from grabbing the railing to steady yourself. Maybe you forgot to turn on the light over the staircase. Maybe your eyesight has been playing tricks on you, but you've put off seeing an eye doctor and getting proper glasses. There are probably dozens of other "maybes" that boil down to you being the cause of your own fall.

On the other hand, maybe there was someone else involved: someone left something on the step, or someone failed to replace the bulb in the stairwell. There could even be a combination of causes: You were in a hurry and didn't turn on the light, so you didn't see the toy that someone else left there.

Accidents on the job don't "just happen," either. They are caused by the actions or inactions of one or more people.

Now for the good news. Just as people cause accidents to happen, they can prevent them from happening. That is the reason for the safe work practices we have established and the posted list of safety rules. It is why employers have regular training sessions to inform and remind you of ways to keep yourselves and your co-workers safe. It's the reason employers provide personal protective equipment that can help keep a potential hazard from causing actual harm.

But no work practices, rules, training, or equipment can prevent an accident from happening. Only you can do that. You follow the checklist, you comply with the safety policies and procedures, you follow the speed limit on your way to the airport – even if you are late, you report safety hazards, you wear your safety glasses, and you wear your hearing protection.

Some of us have special responsibilities that have an effect on everyone's safety. A maintenance supervisor, for example, has to do his or her job correctly or mechanical failures could be followed by accidents. The safety committee chairperson must be sure to post any change in evacuation procedure. And so on. But for the most part, your own safe behavior is your own greatest safeguard. Remember that when you're tempted to take a shortcut or break the safety rule "just this once" or "just for a minute." That one minute could be exactly when the accident doesn't "happen" but is caused.

## Operational Errors and Tactical Errors

*Chapter 7 discussed the James Reason HFACS model. James Reason has written several books relating to operational and tactical errors. He specifically discussed skill-based errors, attention errors, memory errors, technique errors, and decision errors, to name a few. Please review the discussion on errors in Chapter 7.*

## How does brain development affect decision making?

Does the human brain continue to develop with age? When is the brain fully developed? These questions are finally being address with research. Current research conducted by Gargi Talukder, a graduate student in the Neuroscience Program at Stanford University, has turned up some surprises, among them the discovery of striking changes taking place in the brain during the teen years. These findings have altered long-held assumptions about the timing of brain maturation. In key ways, the brain doesn't look like that of an adult until the early 20s.

An understanding of how the brain of an adolescent is changing may help explain a puzzling contradiction of adolescence: young people at this age are close to a lifelong peak of physical health, strength, and mental capacity, and yet, for some, this can be a hazardous age. Mortality rates jump between early and late adolescence. Rates of death by injury between ages 15 to 19 are about six times that of the rate between ages 10 and 14. Crime rates are highest among young males and rates of alcohol abuse are high relative to other ages. Even though most adolescents come through this transitional age well, it is important to understand the risk factors for behavior that can have serious consequences. Genes, childhood experience, and the environment in which a young person reaches adolescence all shape behavior. Adding to this complex picture, research is revealing how all these factors act in the context of a brain that is changing, with its own impact on behavior.

Deborah Yurgelun-Todd and colleagues at the McLean Hospital Brain Imaging Center in Boston, Massachusetts have used functional magnetic resonance imaging to compare the activity of teenage brains to those of adults.

*While adults can use rational processes when facing emotional decisions, teenagers are simply not yet equipped to think through things in the same way.*

The researchers found that when processing emotions, adults have greater activity in their frontal lobes than do teenagers. Adults also have lower activity in their amygdala than teenagers. In fact, as teenagers age into adulthood, the overall focus of brain activity seems to shift from the amygdala to the frontal lobes.

Research now points to the frontal lobes as the place where high-level cognitive decisions about right and wrong, as well as cause-effect relationships are processed. The frontal lobes of the brain have also been implicated in behavioral inhibition, the ability to control emotions and impulses. In contrast, the amygdala is part of the limbic system of the brain and is involved in instinctive "gut" reactions, including "fight or flight" responses. Lower activity in the frontal lobe could lead to poor control over behavior and emotions, while an overactive amygdala may be associated with high levels of emotional arousal and reactionary decision-making.

What does this mean? High-level cognitive decision-making improves as your brain ages. Poor decisions caused by reactionary decision-making decreases as our brain ages. Bottom line: an adult's brain is better suited to tune out emotions and utilize high-level cognitive decision-making.

The results from the McLean study suggest that while adults can to use rational decision-making processes when facing emotional decisions, adolescent brains are simply not yet equipped to think through things in the same way. For example, when deciding whether to ride in a car driven by a drunk friend, an adult can usually put aside her desire to conform and is more likely to make the rational decision against drunk driving. However, a teenager's immature frontal lobes may not be capable of such a coolly rational approach, and the emotional feelings of friendship may be likely to win the battle. As Dr. Yurgelun-Todd told U.S. News, "Good judgment is learned, but you can't learn it if you don't have the necessary hardware."

Jay Giedd and his colleagues at the National Institutes of Mental Health (NIMH) have reached similar conclusions using a brain imaging technique that looks at brain structure rather than activity. Giedd's results suggest that development in the frontal lobe continues throughout adolescence and well into the early twenties. The researchers found that the number of neurons in the frontal lobe continued to increase throughout childhood until an average age of 12.1 years for men and 10.2 years for women. Scientists previously thought that gray matter production and development only occurred during the first 18 months of life. The fact changes are still occurring in the brain during adolescence provides some evidence against some popular theories that suggest that our brains are hardwired during early childhood. These brain imaging studies instead suggest that adolescence may provide a sort of "second chance" to refine behavioral control and rational decision making.

These studies may offer some hope to teenagers suffering from behavioral or emotional problems. The fact that the decision-making centers of the brain continue to develop well into the early twenties could mean that troubled teenagers still have the time as well as the physiology to learn how to control their impulsive behaviors.

The results from these studies do not mean that a teenager will always make irrational decisions. They do, however, suggest that teenagers need guidance as their brains develop, especially in the realm of controlling emotional impulses in order to make rational decisions. It is becoming clear that the adolescent brain is a work in progress, and that parents and educators can help this progress along through open communication and clear boundaries.

## Alcohol and the Teen Brain

Adults drink more frequently than teens, but when teens drink they tend to drink larger quantities than adults. There is evidence to suggest that the adolescent brain responds to alcohol differently than the adult brain, perhaps helping to explain the elevated risk of binge drinking in youth. Drinking in youth, and intense drinking are both risk factors for later alcohol dependence. Findings on the developing brain should help clarify the role of the changing brain in youthful drinking, and the relationship between youth drinking and the risk of addiction later in life.

## Can good decision making be taught?

Research in this area prompted the Federal Aviation Administration (FAA) to produce training directed at improving the decision-making of pilots and led to current FAA regulations that require that decision-making be taught as part of the pilot training curriculum. ADM research, development, and testing culminated in 1987 with the publication of six manuals oriented to the decision-making needs of variously rated pilots. These manuals provided multifaceted materials designed to reduce the number of decision related accidents. The effectiveness of these materials was validated in independent studies where student pilots received such training in conjunction with the standard flying curriculum. When tested, the pilots who had received ADM training made fewer in flight errors than those who had not received ADM training. The differences were statistically significant and ranged from about 10 to 50 percent fewer judgment errors. In the operational environment, an operator flying about 400,000 hours annually demonstrated a 54 percent reduction in accident rate after using these materials for recurrent training. Contrary to popular opinion, good judgment can be taught. Tradition held that good judgment was a natural by-product of experience, but as pilots continued to log accident-free flight hours, a corresponding increase of good judgment was assumed. Building upon the foundation of conventional decision-making, ADM enhances the process to decrease the probability of human error and increase the probability of a safe flight. ADM provides a structured, systematic approach to analyzing changes that occur during a flight and how these changes might affect a flight's safe outcome. The ADM process addresses all aspects of decision-making in the flight deck and identifies the steps involved in good decision-making.

### Steps for good decision-making are:

- Identifying personal attitudes hazardous to safe flight.
- Learning behavior modification techniques.
- Learning how to recognize and cope with stress.
- Developing risk assessment skills.
- Using all resources.
- Evaluating the effectiveness of one's ADM skills.

## Risk Management and ADM

Risk management is an important component of ADM. When a pilot follows good decision-making practices, the inherent risk in a flight is reduced or even eliminated. The ability to make good decisions is based upon direct or indirect experience and education. Consider automotive seat belt use. In just two decades, seat belt use has become the norm, placing those who do not wear seat belts outside the norm, but this group may learn to wear a seat belt by either direct or indirect experience. For example, a driver learns through direct experience about the value of wearing a seat belt when he or she is involved in a car accident that leads to a personal injury. An indirect learning experience occurs when a loved one is injured during a car accident because he or she failed to wear a seat belt. While poor decision-making in everyday life does not always lead to tragedy, the margin for error in aviation is thin. Since ADM enhances management of an aeronautical environment, all pilots should become familiar with and employ ADM.

## Hazard and Risk

Two defining elements of ADM are hazard and risk. A hazard is a real or perceived condition, event, or circumstance that a pilot encounters. When faced with a hazard, the pilot makes an assessment of that hazard based upon various factors. The pilot assigns a value to the potential impact of the hazard, which qualifies the pilot's assessment of the hazard-based risk. Therefore, risk is an assessment of the single or cumulative hazard facing a pilot; however, different pilots see hazards differently. For example, the pilot arrives to preflight and discovers a small, blunt-type nick in the leading edge at the middle of the aircraft's prop. Since the aircraft is parked on the tarmac, the nick was probably caused by another aircraft's prop-wash blowing some type of debris into the propeller. The nick is the hazard (a present condition). The risk is prop fracture if the engine is operated with damage to a prop blade. The seasoned pilot may see the nick as a low risk. He realizes this type of nick diffuses stress over a large area, is located in the strongest portion of the propeller, and based on experience, he doesn't expect it to propagate a crack which can lead to high risk problems. He does not cancel his flight. The inexperienced pilot may see the nick as a high risk factor because he is unsure of the affect the nick will have on the prop's operation and he has been told that damage to a prop could cause a catastrophic failure. This assessment leads him to cancel his flight. Therefore, elements or factors affecting individuals are different and profoundly impact decision-making. These are called human factors and can transcend education, experience, health, physiological aspects, etc. Another example of risk assessment was the flight of a Beechcraft King Air equipped with deicing and anti-icing. The pilot deliberately flew into moderate to severe icing conditions while ducking under cloud cover. A prudent pilot would assess the risk as high and beyond the capabilities of the aircraft, yet this pilot did the opposite. Why did the pilot take this action? Past experience prompted the action. The pilot had successfully flown into these conditions repeatedly although the icing conditions were previously forecast 2,000 feet above the surface. This time, the conditions were forecast from the surface. Since the pilot was in a hurry and failed to factor in the difference between the forecast altitudes, he assigned a low risk to the hazard and took a chance. He and the passengers died from a poor risk assessment of the situation.

## Hazardous Attitudes and Antidotes

Being fit to fly depends on more than just a pilot's physical condition and recent experience. For example, attitude will affect the quality of decisions. Attitude is a motivational predisposition to respond to people, situations, or events in a given manner. Studies have identified five hazardous attitudes that can interfere with the ability to make sound decisions and exercise authority properly: anti-authority, impulsivity, invulnerability, macho, and resignation. (Figure 2.2)

Hazardous attitudes contribute to poor pilot judgment but can be effectively counteracted by redirecting the hazardous attitude so that correct action can be taken.

## Apply the proper antidote

Recognition of hazardous thoughts is the first step toward neutralizing them. After recognizing a thought as hazardous, the pilot should label it as hazardous, then state the corresponding antidote. Antidotes should be memorized for each of the hazardous attitudes so they automatically come to mind when needed. (Figure 2.3)

## Risk

During each flight, the single pilot makes many decisions under hazardous conditions. To fly safely, the pilot needs to assess the degree of risk and determine the best course of action to mitigate risk.

## Assessing Risk

For the single pilot, assessing risk is not as simple as it sounds. For example, the pilot acts as his or her own quality control in making decisions. If a fatigued pilot who has flown 16 hours is asked if he or she is too tired to continue flying, the answer may be no. Most pilots are goal oriented and when asked to accept a flight, there is a tendency to deny personal limitations while adding weight to issues not germane to the mission. For example, pilots of helicopter emergency services (EMS) have been known (more than other groups) to make flight decisions that add significant weight

| The Five Hazardous Attitudes | Antidote |
|---|---|
| **Anti-authority: "Don't tell me."**<br>This attitude is found in people who do not like anyone telling them what to do. In a sense, they are saying, "No one can tell me what to do." They may be resentful of having someone tell them what to do or may regard rules, regulations, and procedures as silly or unnecessary. However, it is always your prerogative to question authority if you feel it is in error. | Follow the rules. They are usually right. |
| **Impulsivity: "Do it quickly."**<br>This is the attitude of people who frequently feel the need to do something, anything, immediately. They do not stop to think about what they are about to do, they do not select the best alternative, and they do the first thing that comes to mind. | Not so fast. Think first. |
| **Invulnerability: "It won't happen to me."**<br>Many people falsely believe that accidents happen to others, but never to them. They know accidents can happen, and they know that anyone can be affected. However, they never really feel or believe that they will be personally involved. Pilots who think this way are more likely to take chances and increase risk. | It could happen to me. |
| **Macho: "I can do it."**<br>Pilots who are always trying to prove that they are better than anyone else think, "I can do it—I'll show them." Pilots with this type of attitude will try to prove themselves by taking risks in order to impress others. While this pattern is thought to be a male characteristic, women are equally susceptible. | Taking chances is foolish. |
| **Resignation: "What's the use?"**<br>Pilots who think, "What's the use?" do not see themselves as being able to make a great deal of difference in what happens to them. When things go well, the pilot is apt to think that it is good luck. When things go badly, the pilot may feel that someone is out to get them or attribute it to bad luck. The pilot will leave the action to others, for better or worse. Sometimes, such pilots will even go along with unreasonable requests just to be a "nice guy." | I'm not helpless. I can make a difference. |

**Figure 2.2** Hazardous Attitudes

*Source:* FAA.gov

| Hazardous Attitude | Antidote |
|---|---|
| Antiauthority: Don't tell me. | Follow the rules. They are usually right. |
| Impulsivity: Do something quickly. | Not so fast. Think first. |
| Invulnerability: It won't happen to me. | It could happen to me. |
| Macho: I can do it. | Taking chances is foolish. |
| Resignation: What's the use? | I'm not helpless. I can make a difference. |

**Figure 2.3** Hazardous Attitudes Antidotes

*Source:* FAA.gov

to the patient's welfare. These pilots add weight to intangible factors (the patient in this case) and fail to appropriately quantify actual hazards such as fatigue or weather when making flight decisions. The single pilot who has no other crew member for consultation must wrestle with the intangible factors that draw one into a hazardous position. Therefore, he or she has a greater vulnerability than a Examining National Transportation Safety Board (NTSB) reports and other accident research can help a pilot learn to assess risk more effectively. For example, the accident rate during night VFR decreases by nearly 50 percent once a pilot obtains 100 hours, and continues to decrease until the 1,000 hour level. The data suggest that for the first 500 hours, pilots flying VFR at night might want to establish higher personal limitations than are required by the regulations and, if applicable, apply instrument flying skills in this environment. full crew. Several risk assessment models are available to assist in the process of assessing risk. The models, all taking slightly different approaches, seek a common goal of assessing risk in an objective manner. Two are illustrated below. The most basic tool is the risk matrix. (Figure 2.4) It assesses two items: the likelihood of an event occurring and the consequence of that event.

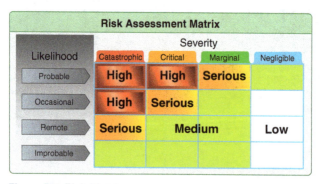

**Figure 2.4** Risk Matrix

*Source:* FAA.gov

## Likelihood of an Event

Likelihood is nothing more than taking a situation and determining the probability of its occurrence. It is rated as probable, occasional, remote, or improbable. For example, a pilot is flying from point A to point B (50 miles) in marginal visual flight rules (MVFR) conditions. The likelihood of encountering potential instrument meteorological conditions (IMC) is the first question the pilot needs to answer. The experiences of other pilots coupled with the forecast, might cause the pilot to assign "occasional" to determine the probability of encountering IMC. The following are guidelines for making assignments.

- Probable – an event will occur several times.
- Occasional – an event will probably occur sometime.
- Remote – an event is unlikely to occur, but is possible.
- Improbable – an event is highly unlikely to occur.

## Severity of an Event

The next element is the severity or consequence of a pilot's action(s). It can relate to injury and/or damage. If the individual in the example above is not an instrument flight rules (IFR) pilot, what are the consequences of him or her encountering inadvertent IMC conditions? In this case, because the pilot is not IFR rated, the consequences are catastrophic. The following are guidelines for this assignment.

- Catastrophic – results in fatalities, total loss
- Critical – severe injury, major damage
- Marginal – minor injury, minor damage
- Negligible – less than minor injury, less than minor system damage

Simply connecting the two factors as shown in Figure 2.4 indicates the risk is high and the pilot must either not fly, or fly only after finding ways to mitigate, eliminate, or control the risk. Although the matrix in Figure 2.4 provides a general viewpoint of a generic situation, a more comprehensive program can be made that is tailored to a pilot's flying. (Figure 2.5)

This program includes a wide array of aviation related activities specific to the pilot and assesses health, fatigue, weather, capabilities, etc. The scores are added and the overall score falls into various ranges, with the range representative of actions that a pilot imposes upon himself or herself.

## Mitigating Risk

Risk assessment is only part of the equation. After determining the level of risk, the pilot needs to mitigate the risk. For example, the pilot flying from point A to point B (50 miles) in MVFR conditions has several ways to reduce risk:

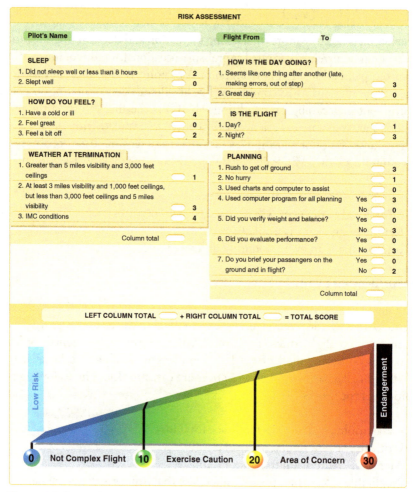

**Figure 2.5** Example of a More Detailed Risk Assessment
*Source:* FAA.gov

- Wait for the weather to improve to good visual flight rules (VFR) conditions.
- Take a pilot who is certified as an IFR pilot.
- Delay the flight.
- Cancel the flight.
- Drive.

One of the best ways single pilots can mitigate risk is to use the IMSAFE checklist to determine physical and mental readiness for flying:

- Illness – Am I sick? Illness is an obvious pilot risk.
- Medication – Am I taking any medicines that might affect my judgment or make me drowsy?
- Stress – Am I under psychological pressure from the job? Do I have money, health, or family problems? Stress causes concentration and performance problems. While the regulations list medical conditions that require grounding, stress is not among them. The pilot should consider the effects of stress on performance.
- Alcohol – Have I been drinking within 8 hours? Within 24 hours? As little as one ounce of liquor, one bottle of beer, or four ounces of wine can impair flying skills. Alcohol also renders a pilot more susceptible to disorientation and hypoxia.
- Fatigue – Am I tired and not adequately rested? Fatigue continues to be one of the most insidious hazards to flight safety, as it may not be apparent to a pilot until serious errors are made.
- Eating – Have I eaten enough of the proper foods to keep adequately nourished during the entire flight?

## The PAVE Checklist

Another way to mitigate risk is to perceive hazards. By incorporating the PAVE checklist into preflight planning, the pilot divides the risks of flight into four categories: **P**ilot-in-command (PIC), **A**ircraft, en**V**ironment, and **E**xternal pressures (PAVE) which form part of a pilot's decision-making process. (Figure 2.6)

With the PAVE checklist, pilots have a simple way to remember each category to examine for risk prior to each flight. Once a pilot identifies the risks of a flight, he or she needs to decide whether the risk or combination of risks can be managed safely and successfully. If not, make the decision to cancel the flight. If the pilot decides to continue with the flight, he or she should develop strategies to mitigate the risks. One way a pilot can control the risks is to set personal minimums for items in each risk category. These are limits unique to that individual pilot's current level of experience and proficiency. For example, the aircraft may have a maximum crosswind component of 15 knots listed in the aircraft flight manual (AFM), and the pilot has experience with 10 knots of direct crosswind. It could be unsafe to exceed a 10 knots crosswind component without additional training. Therefore, the 10 knots crosswind experience level is that pilot's personal limitation until additional training with a certificated flight instructor (CFI) provides the pilot with additional experience for flying in crosswinds that exceed 10

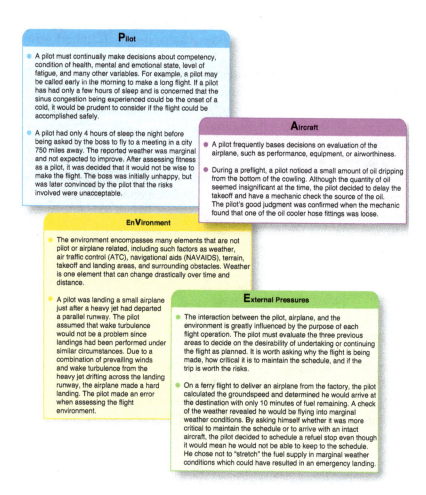

**Figure 2.6** PAVE Checklist

*Source:* FAA.gov

knots. One of the most important concepts that safe pilots understand is the difference between what is "legal" in terms of the regulations, and what is "smart" or "safe" in terms of pilot experience and proficiency.

## P = Pilot in Command (PIC)

The pilot is one of the risk factors in a flight. The pilot must ask, "Am I ready for this trip?" in terms of experience, recency, currency, physical and emotional condition. The IMSAFE checklist provides the answers.

## A = Aircraft

What limitations will the aircraft impose upon the trip? Ask the following questions:

- Is this the right aircraft for the flight?
- Am I familiar with and current in this aircraft? Aircraft performance figures and the AFM are based on a brand-new aircraft flown by a professional test pilot. Keep that in mind while assessing personal and aircraft performance.
- Is this aircraft equipped for the flight? Instruments? Lights? Navigation and communication equipment adequate?
- Can this aircraft use the runways available for the trip with an adequate margin of safety under the conditions to be flown?
- Can this aircraft carry the planned load?
- Can this aircraft operate at the altitudes needed for the trip?
- Does this aircraft have sufficient fuel capacity, with reserves, for trip legs planned?
- Does the fuel quantity delivered match the fuel quantity ordered?

## V = EnVironment

*Weather* Weather is a major environmental consideration. Earlier it was suggested pilots set their own personal minimums, especially when it comes to weather. As pilots evaluate the weather for a particular flight, they should consider the following:

- What is the current ceiling and visibility? In mountainous terrain, consider having higher minimums for ceiling and visibility, particularly if the terrain is unfamiliar.
- Consider the possibility that the weather may be different than forecast. Have alternative plans and be ready and willing to divert, should an unexpected change occur.
- Consider the winds at the airports being used and the strength of the crosswind component.
- If flying in mountainous terrain, consider whether there are strong winds aloft. Strong winds in mountainous terrain can cause severe turbulence and downdrafts and be very hazardous for aircraft even when there is no other significant weather.
- Are there any thunderstorms present or forecast?

- If there are clouds, is there any icing, current or forecast? What is the temperature/dew point spread and the current temperature at altitude? Can descent be made safely all along the route?
- If icing conditions are encountered, is the pilot experienced at operating the aircraft's deicing or anti-icing equipment? Is this equipment in good condition and functional? For what icing conditions is the aircraft rated, if any?

## Terrain

Evaluation of terrain is another important component of analyzing the flight environment.

- To avoid terrain and obstacles, especially at night or in low visibility, determine safe altitudes in advance by using the altitudes shown on VFR and IFR charts during preflight planning.
- Use maximum elevation figures (MEFs) and other easily obtainable data to minimize chances of an inflight collision with terrain or obstacles.

## Airport

- What lights are available at the destination and alternate airports? VASI/PAPI or ILS glideslope guidance? Is the terminal airport equipped with them? Are they working? Will the pilot need to use the radio to activate the airport lights?
- Check the Notices to Airmen (NOTAMS) for closed runways or airports. Look for runway or beacon lights out, nearby towers, etc.
- Choose the flight route wisely. An engine failure gives the nearby airports supreme importance.
- Are there shorter or obstructed fields at the destination and/or alternate airports?

## Airspace

- If the trip is over remote areas, are appropriate clothing, water, and survival gear onboard in the event of a forced landing?
- If the trip includes flying over water or unpopulated areas with the chance of losing visual reference to the horizon, the pilot must be prepared to fly IFR.
- Check the airspace and any temporary flight restriction (TFRs) along the route of flight.

### Nighttime

Night flying requires special consideration.

- If the trip includes flying at night over water or unpopulated areas with the chance of losing visual reference to the horizon, the pilot must be prepared to fly IFR.
- Will the flight conditions allow a safe emergency landing at night?
- Preflight all aircraft lights, interior and exterior, for a night flight. Carry at least two flashlights – one for exterior preflight and a smaller one that can be dimmed and kept nearby for use inside the cabin.

### E = External Pressures

External pressures are influences external to the flight that create a sense of pressure to complete a flight – often at the expense of safety. Factors that can be external pressures include the following:

- Someone waiting at the airport for the flight's arrival.
- A passenger the pilot does not want to disappoint.
- The desire to demonstrate pilot qualifications. • The desire to impress someone. (Probably the two most dangerous words in aviation are "Watch this!")
- The desire to satisfy a specific personal goal ("get-home-itis," "get-there-itis," and "let's-go-itis").
- The pilot's general goal-completion orientation.
- Emotional pressure associated with acknowledging that skill and experience levels may be lower than a pilot would like them to be. Pride can be a powerful external factor!

### Managing External Pressures

Management of external pressure is the single most important key to risk management because it is the one risk factor category that can cause a pilot to ignore all the other risk factors. External pressures put time-related pressure on the pilot and figure into a majority of accidents.

The use of personal standard operating procedures (SOPs) is one way to manage external pressures. The goal is to supply a release for the external pressures of a flight. These procedures include but are not limited to:

- Allow time on a trip for an extra fuel stop or to make an unexpected landing because of weather.

- Have alternate plans for a late arrival or make backup airline reservations for must-be-there trips.
- For really important trips, plan to leave early enough so that there would still be time to drive to the destination.
- Advise those who are waiting at the destination that the arrival may be delayed. Know how to notify them when delays are encountered.
- Manage passengers' expectations. Make sure passengers know that they might not arrive on a firm schedule, and if they must arrive by a certain time, they should make alternative plans.
- Eliminate pressure to return home, even on a casual day flight, by carrying a small overnight kit containing prescriptions, contact lens solutions, toiletries, or other necessities on every flight.

The key to managing external pressure is to be ready for and accept delays. Remember that people get delayed when traveling on airlines, driving a car, or taking a bus. The pilot's goal is to manage risk, not create hazards (Figure 2.6).

## The Decision-Making Process

An understanding of the decision-making process provides the pilot with a foundation for developing ADM and SRM skills. While some situations, such as engine failure, require an immediate pilot response using established procedures, there is usually time during a flight to analyze any changes that occur, gather information, and assess risk before reaching a decision.

Risk management and risk intervention is much more than the simple definitions of the terms might suggest. Risk management and risk intervention are decision-making processes designed to systematically identify hazards, assess the degree of risk, and determine the best course of action. These processes involve the identification of hazards, followed by assessments of the risks, analysis of the controls, making control decisions, using the controls, and monitoring the results.

## Chapter Questions

1. Research now points to the frontal lobes as the place where_____ are processed
   A) high level cognitive decisions
   B) emotional reactions
   C) Both A and B

2. New research has shown that an overactive amygdala may be associated with _____
   A) high levels of emotional arousal and reactionary decision-making.
   B) High-level cognitive decision-making abilities
   C) Both A and B
3. Explain the 5 hazardous attitudes.
4. Explain the antidotes for the 5 hazardous attitudes.
5. How would you define a hazard?
6. How would you define a risk?

# Hazardous Attitudes

1. Identify three mental processes of flight.
2. Learn the five basic hazardous attitudes.
3. Know the ten steps that could lead to disaster.
4. Know who can be affected by a hazardous attitude.
5. Name three air carrier accidents that included an NTSB probable cause that was due to a hazardous attitude.

Does attitude make a difference in how we do things, or complete tasks? If we say yes to that question, how big of a difference does it actually make? Most people would agree that the answer is yes, and that attitude does make a big difference in how a task is actually accomplished. In aviation, attitude can make a tremendous difference in the outcome of a particular event. This goes for everyone involved in aviation. If the safety of a particular flight relies on everyone involved, from the pilot to the air traffic controller, then one person's hazardous attitude can make all the difference. The Aircraft Owners and Pilots Association (AOPA) states that the key to good aeronautical decision making is an understanding of hazardous attitudes.

For pilots, judgement is the ability to make a decision that provides the greatest flight safety. Information needs to be continually analyzed about the aircraft, the environment, and yourself through mental processes. This involves three mental processes in flight:

- Automatic Reaction: a rote level, motor skill response to normal or emergency situation. This should be an ingrained habit pattern response.
- Problem Resolving: time permitting, we

- Systematically analyze the situation.
- Decide on the course of action that will provide the greatest flight safety.
- Repeated Reviewing: constant review of the situation, including changes as they occur, until the flight is completed.

When looking at judgement and the three mental processes in flight, it is important to remember that how the pilot handles responsibility depends on his/her attitude toward safety, themselves, and flying. Remember, attitudes are learned.

So, we know that judgement is important to pilot decision making. Let's take a look at the hazardous attitudes that can affect pilot judgement. There are five basic hazardous attitudes:

- Anti-Authority
- Impulsivity
- Invulnerability
- Macho
- Resignation

Let's take a closer look at each of these and how we can possibly solve each of them. We will refer to these as antidotes.

- Anti-Authority: Don't tell me what to do!
  Solution: Follow the rules. They are usually right.
- Impulsivity: Do something – Quick!
  Solution: Not so fast, think first.
- Invulnerability: It won't happen to me.
  Solution: It could happen to me. Think about the worst possible accident; it could happen to you.
- Macho: I can do it.
  Solution: Taking chances is foolish.
- Resignation: What's the use?
  Solution: I'm not helpless. I can make a difference.

Now that we have looked at judgement and the attitudes that can have an effect on it, let's take a look at ten additional steps (attitudes) that could lead to disaster.

- Ambiguity:
  - Two or more independent sources of information do not agree.
- Fixation or Preoccupation:
  - Attention of the crew is focused on one item, event, or condition to the exclusion of all other activity in the cockpit.

- Empty Feeling or Confusion:
  - A pilot or other crew member is unsure of the state of the aircraft or its conditions.
- Violating Minimums:
  - Minimums are intentionally violated, or consideration is given to doing so.
- Undocumented Procedure:
  - Consideration given to use of an undocumented procedure or when an undocumented procedure is in fact used.
- Nobody Flying the Aircraft:
  - May result from fixation or preoccupation. May also occur in routine flight conditions.

Aircraft accident files are filled with cases where no one was delegated to fly the aircraft. Captain or Pilot-in-Command must specifically delegate this responsibility: "I'll fly the airplane and you take care of the problem."

- Nobody Looking Out the Window (Outside):
  - With use of new sophisticated flight management systems (FMS), the possibility of both pilots being "heads down" at the same time is certainly real.
- Failure to Meet Target (Expectations):
  - Parameters or expectations of events are not met.
  - Examples: Expected fuel burn or anticipated takeoff thrust settings
- Unresolved Discrepancies:
  - Confusion, questions, or statements of concern.
- Departure from Standard Operating Procedure:
  - Standard operating procedure not used at appropriate time.

Hazardous attitudes might seem like an issue that should go away as a pilot gains more experience. Unfortunately, this is not the case. There have been many examples of air carrier accidents that have occurred due to some of the hazardous attitudes we previously discussed. Here are three examples, along with the NTSB's probable or contributing cause:

- Eastern Air Lines Flight 401, Lockheed L-1011 Tristar, 101 fatalities (Figure 3.1)
  - NTSB Probable Cause: "The failure of the fight crew to monitor the flight instruments during the final 4 minutes of flight, and to detect an unexpected descent soon enough to prevent impact with the ground. Preoccupation with a malfunction of the nose landing gear position indicating system distracted the crew's attention from the instruments and allowed the descent to go unnoticed."

**Figure 3.1** Eastern Air Lines Flight 401
*Source:* NTSB.gov

- United Airlines Flight 173, Douglas DC-8, 10 fatalities (Figure 3.2)
  - NTSB Probable Cause: "The failure of the captain to monitor properly the aircraft's fuel state and to properly respond to the low fuel state and the crewmember's advisories regarding fuel state. This resulted in fuel exhaustion to all engines. His inattention resulted from preoccupation with a landing gear malfunction and preparations for a possible landing emergency. Contributing to the accident was the failure of the other two flight crewmembers either to fully comprehend the criticality of the fuel state or to successfully communicate their concern to the captain."

**Figure 3.2** United Air Lines Flight 173
*Source:* NTSB.gov

- Pinnacle Airlines Flight 3701, 2 fatalities (Figure 3.3)
  - NTSB Probable Cause: "The National Transportation Safety Board determines that the probable causes of this accident were (1) the pilots' unprofessional behavior, deviation from standard operating procedures, and poor airmanship, which resulted in an in-flight emergency from which they were unable to recover . . ."

**Figure 3.3** Pinnacle Air Lines Flight 3701

*Source:* Bureau of Accident Investigation

Hazardous attitudes are prevalent throughout aviation. How a pilot handles these attitudes is the key to how well the outcome of a particular flight ends up. These attitudes can affect judgement, and have a real effect on the safety of flight. Although it is easy to see where a hazardous attitude may have affected the outcome of a flight when the end result is an accident, it is tougher to see where a pilot corrected their hazardous attitude and had a different outcome. Although we commonly do not read about these situations, pilots who correct their hazardous attitudes have successfully prevented a human error accident from occurring.

## Chapter Questions

1. Name one of the three mental processes of flight:
   A) Automatic reaction
   B) Thinking
   C) Resignation
   D) All of the above
2. Hazardous attitudes effect:
   A) Pilots
   B) Managers
   C) Air Traffic Controllers
   D) All of the above
3. AOPA believes that hazardous attitudes effect all pilots
   A) True
   B) False

# Managing Fatigue

---

## LEARNING OBJECTIVES

1. Identify what percentage of accidents in aviation are caused by fatigue.
2. Know the definition of fatigue.
3. Name the four types of fatigue.
4. Identify the biggest cause of fatigue.
5. Name the five stages of sleep.
6. Identify other causes of fatigue.
7. Know the definition of circadian rhythm.
8. Identify four accidents related to fatigue.

---

Fatigue is a very prevalent issue in aviation. But, why is this? And, should it be? Fatigue is something that almost everyone can say that they have experienced. This is not an issue that just affects pilots. The problem, of course, in aviation fatigue can have disastrous results. It has been reported that 80% of aircraft accidents are related to fatigue. Beginning in 1990, fatigue awareness and duty time limitations were included in the top ten most wanted safety improvements list that is put out each year by the National Transportation Safety Board (NTSB). In this particular chapter will focus on this issue, with the particular emphasis on the causes and effects of fatigue.

First, let's look at the definition of fatigue. As a noun, the definition according to Merriam-Webster: "the state of being very tired: extreme weariness." As a verb, Merriam-Webster defines fatigue as: "to make (someone) tired."

Those definition are the what the dictionary defines fatigue as, but what is it really? While awake, the healthy, well-nourished and rested human brain is capable of prodigious feats of sensory perception, symbol manipulation, logic, analytic thought, language, and problem-solving. However, because of its biologic nature, the brain cannot run continuously in the awake conscious mode, but requires scheduled maintenance and recharge cycles for efficient function. The awake, functioning brain seems to deplete neurons and biochemical capability, build up toxins and metabolic by-products, and starts to run down. This "running down" is manifest as declines in mental performance, judgment, and complex decision- making, and is associated with a variety of symptoms we commonly experience as, and refer to, "fatigue."

In aviation, we classify fatigue into four main types:

- <u>Mental Fatigue</u>: a transient decrease in maximal cognitive performance resulting from prolonged periods of cognitive activity.
- <u>Physical Fatigue</u>: the transient inability of a muscle to maintain optimal physical performance, and is made more severe by intense physical exercise.
- <u>Acute Fatigue</u>: a sudden onset of physical and mental exhaustion or weariness, particularly after a period of mental or physical stress. This type of fatigue is usually short term and is considered normal.
- <u>Chronic Fatigue</u>: Long-continued fatigue not relieved by rest. It is indicative of disease such as tuberculosis, diabetes, or other conditions of altered body metabolism. Chronic fatigue can either be psychological or emotional.

The primary cause of fatigue is one that people usually have the most control over. This factor is lack of sleep. On average, human adults physiologically require about 8 hours of sleep. Studies show that the average range is 6 to 8 hours. There are five stages of sleep:

- Stage 1 is the transition between consciousness and sleep. You can generally hear and respond to someone.
- Stage 2 is a light sleep. You are easily awakened but you're not aware of your surroundings.
- Stages 3 and 4 are deep slumber – this is a very restorative phase.
- Stage 5 is known as REM or rapid eye movement sleep, and it's the stage of sleep where you dream. Researchers believe your eyes move at this stage of sleep because you're scanning the images in your dreams. It's thought to be important for learning and consolidation of memory.

A complete sleep cycle can last between 60 and 90 minutes, with a typical sleep moving through the cycle several times, but each cycle will vary in length. Whenever you're sleep deprived, your body will try first to catch up on deep sleep (Stages 3 and 4) and REM sleep (Stage 5). The human body is built for 16 hours awake and 8 hours of sleep. So, sleep loss can be acute (within 24 hours) or cumulative (over several days). Sleep loss that is cumulative is commonly known as "sleep debt." Sleep debt is fairly easy to figure out. As stated earlier, an average adult has a sleep requirement of eight hours. If a person actually sleeps six hours, the sleep debt is now two hours. So, if we have four nights of six hours of sleep, we now have a sleep debt of 8 hours. In effect, we have missed a whole night's sleep in that four day period, as sleep debt is cumulative.

Researchers from NASA have shown performance & alertness can be maintained up to 12 hours of wakefulness. But:

- At 16 hours, studies have shown progressive accident/injury rate at 3 times that a 9 hours.
- At 17 hours studies have shown performance levels consistent with .04% blood/alcohol level, the equivalent of drinking 2 beers.
- At 22 hours studies have shown performance levels consistent with .08% blood/alcohol level, the equivalent of drinking 5 beers.

In response to the accident of Colgan Air flight 3407, which occurred in February 2009, the FAA issued new crew rest and duty limitations under Federal Aviation Regulation (FAR) Part 117. This policy applies to most passenger operations, but does not apply to cargo operations. FAR Part 117 can be read here: https://www.faa.gov/regulations_policies/rulemaking/recently_published/media/2120-AJ58-FinalRule.pdf

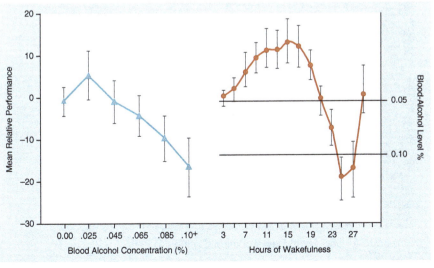

Lack of sleep is just one cause of fatigue. Other factors or causes of fatigue include:

- Dehydration
- Noise & Vibration
- Illness
- OTC Medications
- Impaired Vision
- Lack of Restful Sleep
- Sleep disorders (sleep apnea)
- Food Allergies
- Diabetes
- Depression

Another aspect of fatigue is circadian rhythm. These are physical, mental and behavioral changes that follow a roughly 24-hour cycle, responding primarily to light and darkness in an organism's environment. In essence, this is commonly referred to as our "body clock." It is controlled by brain function and controls timing of physiological activity (alertness, performance, etc.). The lowest point in a normal cycle is from 12:00 am–6:00 am, specifically 3:00 am–5:00 am.

Below are some examples of accidents or incidents that are known to have at least a partial cause from fatigue:

1. **Corporate Airlines Flight 5966**
   Date: October 19, 2004 Aircraft: BAE-J32
   Location: Kirksville, Missouri
   Accident Description: Aircraft struck trees on final approach. Fatalities: 11
   NTSB quote: ***"Fatigue contributed to pilots degraded performance."***

Corporate Airlines Flight 5966 Accident Scene
*Source:* NTSB.gov

2. **Shuttle America Flight 6448**
   Date: February 18, 2007 Aircraft: EMB 175
   Location: Cleveland, Ohio
   Accident Description: Over ran end of runway. Fatalities: none
   NTSB quote: *"Captains fatigue contributed to accident."*

Shuttle America Flight 6448 Accident Scene
*Source:* NTSB.gov

3. **Go! Airlines Flight 1002**
   Date: February 21, 2008
   Aircraft: CRJ-200
   Location: Hilo, Hawaii
   Incident description: Both crew members fell asleep between Honolulu and Hilo.
   NTSB quote: "The incident pilots' lack of adequate sleep, together with the low workload associated with the cruise phase of the flight, likely contributed to the pilots inadvertently falling asleep."
   Some of the additional information from the NTSB report highlights the fatigue related issues with this particular flight:
   "...the day of the incident was the third consecutive day that both pilots started duty at 0540. This likely caused the pilots to receive less daily sleep than is needed to sustain optimal alertness and resulted in an accumulation of sleep debt and increased levels of daytime fatigue. The first officer stated he needed between 7.5 and 8 hours of sleep per night to feel rested. He estimated that he had spent about 7 hours 25 minutes in bed the night before the incident, and about 6 hours 55 minutes in bed during each of the previous two nights. Thus, the first officer's self- reported sleep history indicated an accumulated sleep debt of between 1 hour 15 minutes and 2 hours 45 minutes in the 72 hours before the incident."

4. **U.S. Air Force flight**

Date: July 20, 2012

Aircraft: Boeing C-17A Globemaster III

Location: Tampa, Florida

Incident Description: The aircraft, bound for McDill Air Force Base in Florida, mistakenly landed at Peter O. Knight Airport, also in Florida. Air Force Investigation quote: "flew into complex airfields, dealt with multiple mission changes and flew long mission legs with several stops each day."

Some of the additional information from the Air Force report highlights the fatigue related issues with this particular flight:

". . . the pilot was acting at a 79 percent cognitive effectiveness and the copilot 89 percent. In comparison, a 0.08 percent blood alcohol level reduces the cognitive capacity to 70 percent."

## Chapter Questions

1. Fatigue can be caused by:
   A) Lack of sleep
   B) Dehydration
   C) Depression
   D) All of the above
2. The four types of fatigue include:
   A) Sleep Fatigue
   B) Intellectual Fatigue
   C) Mental Fatigue
   D) None of the above
3. Circadian rhythm is commonly known as our body clock.
   A) True
   B) False

# Mid Air Collision Avoidance

**5**

## LEARNING OBJECTIVES

1. Know the profile of midair collisions.
2. Know what causes a midair collision.
3. Explain the limitations of the eye.
4. Know how to scan for traffic.
5. Familiarize yourself with the collision avoidance checklist.

## Introduction

By definition and function, the human eye is one of the most important and complex systems in the world. Basically, its job is to accept images from the outside world and transmit them to the brain for recognition and storage. In other words, the organ of vision is our prime means of identifying and relating to what is going on around us.

It has been estimated that 80 percent of our total information intake is through the eyes. In the air, we depend on our eyes to provide most of the basic input necessary for performing during a flight: attitude, speed, direction, and proximity to things (like the ground), and opposing air traffic that may constitute a danger of in-flight collision. As air traffic density and aircraft closing speeds increase, the problems of in-flight collision grows proportionately, and so does the importance of the "eyeball system." A basic understanding of the eyes' limitation in target detection is probably the best insurance a pilot can have against running into another airplane – something that can spoil your whole day.

## Profile of Midair Collisions

Studies of midair collisions reveal certain definite warning patterns. It may be surprising to learn that nearly all midair collisions occur during daylight hours and in VFR conditions. Perhaps not so surprising is that the majority happen within five miles of an airport, in the areas of greatest traffic concentration, and usually on warm weekend afternoons when more pilots are doing more flying.

## Not What You Might Expect

Also surprising, perhaps, is the fact that the closing speed (rate at which two aircraft come together) is relatively slow, usually much slower than the airspeed of either aircraft. In fact, the majority of in-flight collisions are the result of a faster aircraft overtaking and hitting a slower plane.

Statistics on 105 in-flight collisions that occurred from 1964 to 1968 show that 82 percent had convergence angles associated with one aircraft overtaking another. Specifically, 35 percent were from 0 to 10 degrees – straight from behind. Only 5 percent were from a head-on angle. These numbers, plus the fact that 77 percent occurred at or below 3,000 feet (with 49 percent at or below 500 feet) imply accurately that in-flight collisions generally occur in the traffic pattern and primarily on final approach. Collisions occurring enroute generally are at or below 8,000 feet and within 25 miles of an airport.

## No Pilot is Immune

The pilots involved in such mishaps ranged in experience from first solo to 15,000 hours, and their reasons for flying on the accident day were equally varied. Some examples:

- A 19-year-old private pilot flying a VFR cross-country in a Cessna 150 collided with two seasoned airline pilots flying a Convair 580 under IFR control.
- A 7,000-hour commercial pilot on private business in a twin Beech overtook a Cherokee on final, which carried a young flight instructor giving dual to a pre-solo student pilot.
- Two commercial pilots, each with well over 1,000 hours, collided while ferrying a pair of new single-engine aircraft.
- Two private pilots with about 200 hours logged between them collided while on local pleasure flights in Piper Cubs.

There is no way to say whether the inexperienced pilot or the older, more experienced pilot is more likely to be involved in an in-flight collision.

A beginning pilot has so much to think about he or she may forget to look around. On the other hand, the older pilot, having sat through many hours of boring flight without spotting any hazardous traffic, may grow complacent and forget to scan. No pilot is invulnerable.

## Midair Collison Causes

What causes in-flight collisions? Undoubtedly, increasing traffic and higher closing speeds represent potential. For instance, a jet and a light twin have a closing speed of about 750 mph. It takes a minimum of 10 seconds, says the FAA. for a pilot to spot traffic, identify it, realize it is a collision threat, react, and have the aircraft respond. But two planes converging at 750 mph will be less than 10 seconds apart when the pilots are first to detect each other!

These are all causal factors, but the reason most often noted in the statistics reads: "Failure of pilot to see other aircraft," which means that the see-and-avoid system broke down. In most cases, at least one of the pilots involved could have seen the other in time to avoid contact, if he or she had just been using the visual senses properly. In sum, it is really that complex, vulnerable little organ – the human eye- which is the leading cause of inflight collisions.

Let's take a look at how its limitations affect your flight.

### Limitations of the Eye

The eye, and consequently vision, is vulnerable to just about everything: dust; fatigue; emotion; germs; fallen eyelashes; age; optical illusions; and the alcoholic content of last night's party. In flight, vision is altered by atmospheric conditions, windshield distortion, too much (or too little) oxygen, acceleration, glare, heat, lighting, aircraft design and forth.

Most of all, the eye is vulnerable to the vagaries of the mind. We can "see" and identify only what the mind lets us see. For example, a daydreaming pilot staring out into space sees no approaching traffic and is probably the number one candidate for an in-flight collision.

## Accommodation

One function of the eye that is a source of constant problems to the pilot (though he or she is probably never aware of it) is the time required for accommodation. Our eyes automatically accommodate for (or refocus on) near and far objects. But the change from something up close, like a dark panel two feet away, to a well-lighted landmark or aircraft target a mile or so away, takes one to two seconds or longer for eye accommodation. That can be a long time when you consider that you need 10 seconds to avoid in-flight collisions.

## Empty-Field Myopia

Another focusing problem usually occurs at very high altitudes, but it can happen even at lower levels on vague, colorless days above a haze or cloud layer when no distinct horizon is visible. If there is little or nothing to focus on at infinity, we do not focus at all. We experience something known as "empty-field myopia:" we stare, but we see nothing, even opposing traffic, if it should enter our visual field.

## Binocular Vision

The effects of what is called "binocular vision" have been studied seriously by the National Transportation Safety Board (NTSB) during investigations of in-flight collisions, with the conclusions that this too is a casual factor. To actually accept what we see, we need to receive cues from both eyes. If an object is visible to one eye, but hidden from the other by a windshield post or other obstruction, the total image is blurred and not always acceptable to the mind.

## Tunnel Vision

Another inherent eye problem is that of narrow field of vision. Although our eyes accept light rays from an arc of nearly 200 degrees, they are limited to a relatively narrow area (approximately 10–15 degrees) in which they can actually focus and classify an object. Though we can perceive movement in the periphery, we cannot identify what is happening out there, and we tend not to believe what we see out of the corner of our eyes. This, aided by the brain, often leads to "tunnel vision."

## Blossom Effect

This limitation is compounded by the fact that at a distance, an aircraft on a collision course with you will appear to be motionless. It will remain in a seemingly stationary position, without appearing either to move or to grow in size for a relatively long time, and then suddenly bloom into a huge mass filling one of your windows. This is known as "blossom effect." Since we need motion or contrast to attract our eyes' attention, this effect becomes a frightening factor when you realize that a large bug smear or dirty spot on the windshield can hide a converging plane until it is too close to be avoided.

## Environmental Effects

In addition to the built-in problems, the eye is also severely limited by environment. Optical properties of the atmosphere alter the appearance of traffic, particularly on hazy days. "Limited visibility" actually means "limited vision." You may be legally VFR when you have three miles, but at that distance on a hazy day, opposing traffic is not easy to detect. At a range closer than three miles, opposing traffic may be detectable, but no longer avoidable.

Lighting also affects our vision stimuli. Glare, usually worse on a sunny day over a cloud deck or during flight directly into the sun, makes objects hard to see and makes scanning uncomfortable. Also, an object that is well lighted will have a high degree of contrast and will be easy to detect, while one with low contrast at the same distance may be impossible to see. For instance, when the sun is behind you, an opposing aircraft will stand out clearly, but when you are looking into the sun and your traffic is "backlighted," it's a different story.

Another contrast problem area is trying to find an airplane over a cluttered background. If it is between you and terrain that is Varicolored or heavily dotted with buildings, it will blend into the background until it is quite close.

## Human Factors

And, of course, there is the mind, which can distract us to the point of not seeing anything at all, or lull us into cockpit myopia – staring at one instrument without even "seeing" it. How often have you filed IFR on a VFR day, settled back at your assigned altitude with autopilot on, and then never looked outside, feeling secure that "Big Daddy Radar" will protect you from all harm? Don't fall for this trap. Remember, the radar system has its limitations too! It is fine to depend on instruments, but not to the exclusion of the see-and-avoid system, especially on days when there are pilots not under radar surveillance or control flying around in the same sky. Also remember that our Air Traffic Control (ATC) system is not infallible, even when it comes to providing radar separation between aircraft flying on IFR flight plans.

As you can see, visual perception is affected by many factors. It all boils down to the fact that pilots, like anyone else, tend to overestimate their visual abilities and to misunderstand the limitations of their eyes. Since the number one cause of in-flight collisions is the failure to properly adhere to the see-and- avoid concept, we can conclude that the best way to avoid them is to learn how to use our eyes in an efficient external scan.

## How to Scan

What is the perfect scan? There is none, or at least there is no one scan that is best for all pilots. The most important thing is for each pilot to develop a scan that is both comfortable and workable.

The best way to start is by getting rid of bad habits. Naturally, not looking out at all is the poorest scan technique, but glancing out at intervals of five minutes or so is also poor when you remember that it only takes seconds for a disaster to happen. Check yourself the next time you are climbing out, making an approach, or just bouncing along over a long cross-country route. See how long you go without looking out the window.

Glancing out and giving it the once-around without stopping to focus on anything is practically useless. So is staring out into one spot for long periods of time (even though it may be great for meditation).

So much for the bad habits. Learn how to scan properly; first, by knowing where to concentrate your search. It would be preferable, naturally, to look everywhere constantly but, as this technique is not practical, concentrate on the areas most critical to you at any given time. In the traffic pattern especially, clear before every turn, and always watch for traffic making an improper entry into the pattern. On descent and climb-out, make gentle S-turns to see if anyone is in your way. (In addition, make clearing turns before attempting maneuvers, such as pylons and S-turns about a road.)

During the very critical final approach stage, don't forget to look behind and below, at least once; and avoid tunnel vision. Pilots often rivet their eyes to this point of touchdown. You may never arrive at it if another pilot is aiming for the same numbers at the same time!

In normal flight, you can generally avoid the threat of an in-flight collision by scanning and area 60 degrees to the left and to the right of your center visual area. This advice does not mean you should forget the rest of the area you can see from side windows every few scans. Horizontally, the statisticians say, you will be safe if you scan 10 degrees up and down from your flight vector (Figure 5.1). This technique will allow you to spot any aircraft that is at an altitude that might prove hazardous to your own flight path, whether it is level with you, below and climbing, or above and descending.

The slower your plane, the greater your vulnerability, hence the greater scan area required.

But don't forget that your eyes are subject to optical illusions and can play some nasty tricks on you. At one mile, for example, an aircraft flying below your altitude will appear to be above you. As it nears, it will seem to descend and go through your level, yet, all the while it will be straight and level below you. one in-flight collision occurred when the pilot of the higher-flying airplane experienced this illusion and dove his plane right into the path of the aircraft flying below.

Though you may not have much time to avoid another aircraft in your vicinity, use your head when making defensive moves. Even if you must maneuver to avoid a real in-flight collision, consider all the facts. If you miss the other aircraft but stall at a low altitude, the results may be just as bad for you as a collision.

*You can generally avoid the threat of an inflight collision by vertically scanning 60 degrees to the left and right and horizontally scanning 10 degrees up and down*

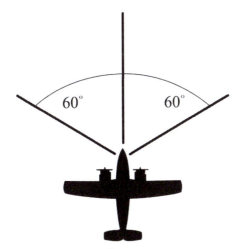

**Figure 5.1** Prioritize Your Scan

*Source:* Michael Ullrich

## Scan Patterns

### *Block System*

Your best defense against in-flight collisions is an efficient scan pattern. Two basic scans that have proved best for most pilots are variations on a technique called the "block" system. This type of scan is based on the theory that traffic detection can be made only through a series of eye fixations at different points in space. Each of these fixes becomes the focal point of your field of vision (a block 10–15° wide). By fixating every 10–15 degrees wide, you should be able to detect any contrasting or moving object in each block. This gives you 9–12 "blocks" in your scan area, each requiring a minimum of one to two seconds for accommodation and detection.

### Side-to-Side Block Scan

One method of block scan is the "side-to-side" motion. Start at the far left of your visual area and make a methodical sweep to the right, pausing in each block to focus. At the end of the scan, return to the panel.

### Front-to-Side Block Scan

The second form is the "front-to-side" version. Start with a fixation in the center block of your visual field (approximately the center of the front windshield in front of the pilot). Move your eyes to the left, focusing in each block, swing quickly back to the center block, and repeat the performance to the right (Figure 5.2).

There are other methods of scanning of course, some of which may be as effective for you as the two preceding types. Unless some series of fixations is made, however, there is little likelihood that you will be able to detect all targets in your scan area. When the head is in motion, vision is blurred and the mind will not register targets as such.

## The Time-sharing Plan

External scanning is just part of the pilot's total eyeball job. To achieve maximum efficiency in flight, one has to establish a good internal (panel) scan as well and learn to give each instrument its proper share of time. The amount of time one spends eyeballing outside the cockpit in relation to what is spent inside depends, to some extent, on the workload inside the cockpit and the density of traffic outside. Generally, the external scan will take about three to four times as long as a look around the instrument panel.

A major company conducted an experimental scan training course, using military pilots ranging in experience from 350 to over 4,000 hours. They discovered that the average time needed to maintain a flight situation *status quo* was three seconds for panel scan and 17 seconds for outside. (Since military pilots are most likely flying a more consistent schedule than most general aviation pilots, we should allow six or seven seconds on the panel.)

## Panel Scan

An efficient instrument scan is good practice, even if you limit your flying to VFR conditions, and being able to quickly scan the panel gives one a better chance of doing an effective job outside as well. The following panel scan system is taught by FAA and AOPA Air Safety Foundation to instrument students (Figure 5.3).

Two scanning methods that have proved to be the most effective for pilots involve the "block" system of scanning, which is based on the thoery that "traffic detection can be made only through a series of eye fixations at different points in space." In application, the viewing area (windshield) is divided into segments, and the pilot methodically scans for traffic in each block of airspace in sequential order.

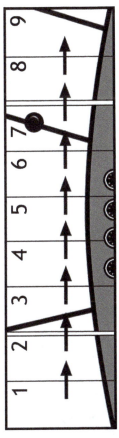

Side-to-side scanning method. Start at the far left of your visual area and make a methodical sweep to the right, pausing in each block of viewing area to focus your eyes. At the end of the scan, return to the panel.

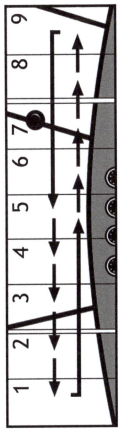

Front-to-side scanning method. Start in the center block of your visual field (center of front windshield); move to the left, focusing in each block, then swing quickly to the center block after reaching the last block on the left and repeat the performance to the right.

*FIGURE 2*

**Figure 5.2** Two Recommended Scanning Methods

*Source:* Michael Ullrich

COLLISION AVOIDANCE CHECKLIST

1. Check Yourself
2. Plan Ahead
3. Clean Windows
4. Adhere to S.O.P's
5. Avoid Crowds
6. Compensate for Design
7. Equip for Safety
8. Talk and Listen
9. SCAN!

*The panel scan shown here involves skimming over the attitude indicator each time your scan moves on to a new instrument. (1) Start with the attitude indicator, then move to the directional gyro for heading; (2) move on to the altimeter; (3) airspeed indicator; (4) rate-of-climb indicator; (5) turn-and-bank indicator, include your VOR (6) and engine instruments (7) every third scan or so, or as the flight situation dictates.*

**Figure 5.3** Panel Scan

*Source:* Michael Ullrich

- Start with the attitude indicator. It will show changes in attitude the two most critical areas of flight: heading and altitude.
- Move to the directional gyro for heading.
- Move to altimeter.
- Check the airspeed indicator.
- Look at rate of climb (VSI).
- Look at the turn and bank indicator (or turn coordinator).

It is a good idea to skim over the attitude indicator each time you move on to a new instrument, as the AI is your chief control instrument. Include your VOR and engine instruments every third scan or so, or as the flight situation dictates.

Developing an efficient time-sharing plan takes a lot of work and practice, but it is just as important as developing good landing techniques. The best way is to start on the ground, in your own airplane or the one you usually fly, and then use your scans in actual practice every chance you get.

## Collision Avoidance Checklist

Collision avoidance involves much more than proper eyeball techniques. You can be the most conscientious scanner in the world and still have an in-flight collision if you neglect other important factors in the overall see-and-avoid picture. It might be helpful to use a collision avoidance checklist as religiously as you do the pre-takeoff and landing lists. Such a checklist might include the following nine items:

## Check Yourself

Start with a check of your own condition. Your eyesight, and consequently your safety, depend on your mental and physical condition.

## Plan Ahead

Plan your flight ahead of time. Have charts folded in proper sequence and with handy reach. Keep your cockpit free of clutter. Be familiar with headings, frequencies, distances, etc., ahead of time; so, that you spend minimum time with your head down in your charts. Some pilots even jot these things down on a flight log before takeoff. Check your maps and the special general and area notices in AIM in advance for restricted areas, oil burner routes, intensive student jet training areas and other high-density spots.

## Clean Windows

During the walk-around, make sure your windshield is clean. If possible, keep all windows clear of obstructions, like solid sun visors and curtains.

## Adhere to Standard Operating Procedures

Stick to Standard Operating Procedures and observe the regulations of flight, such as correct altitudes and proper pattern practices. You can get into big trouble, for instance, by "sneaking" out of your proper altitude as cumulous clouds begin to tower higher and higher below you, or by skimming along the tops of clouds without observing proper separation. Some typical situations involving in-flight mishaps around airports include entering a right-hand pattern at an airport with left-hand traffic; or entering downwind so far ahead of the traffic pattern that you may interfere with traffic taking off and heading out in your direction. In most in-flight collisions, at least one of the pilots involved was not where he was supposed to be.

## Avoid Crowds

Avoid crowded airspace enroute, such as directly over a VOR. You can navigate on VFR days just as accurately by passing slightly to the left or right of the VOR stations. Pass over airports at a safe altitude, being particularly careful within a 25-mile radius of military airports and busy civil fields. Military airports usually have a very high concentration of fast-moving jet traffic in the vicinity and a pattern that extends to 2,500 feet above the surface. Jets in climb-out may be going as fast as 500 mph.

## Compensate for Design

Compensate for your aircraft's design limitations. All planes have blind spots; know where they are in your aircraft. For example, a high wing aircraft that has a wing down in a turn blocks the area you are turning into. A low wing blocks the area beneath you. And one of the most critical midair potential situations is a faster low-wing plane overtaking and descending on a high wing on final approach.

## Equip for Safety

Your airplane can, in fact, help avoid collisions. Certain equipment that was once priced way above the light plane owner's reach, now is available at reasonable cost to all aviation segments. High intensity strobe lights increase your contrast by as much as 10 times day or night and can be installed for about

$200 each. In areas of high density, use your strobes or your rotating beacon constantly, even during daylight hours. The cost is pennies per hour – small price to pay for making your aircraft easier for other pilots to see.

Transponders, available in quick installation kits for under $1,000, significantly increase your safety by allowing radar controllers to keep your traffic away from you and vice versa. Now mandatory for flight into many high density airport areas, transponders also increase your chances of receiving radar traffic advisories, even on VFR flights.

## Speak Up, Listen Up

Use your radio, as well as your eyes, When approaching an airport. If you are operating close enough to the airport in terms of altitude and location to be near traffic going to or from that airport, consider making a call to state your position, altitude and intentions. Find out what the local traffic situation is. At an airport with radar service, call the approach control frequency and let them know where you are and what you are going to do. At non-towered fields, listen to the common traffic advisory frequency (CTAF) to develop a mental picture of traffic around you.

Since detecting a tiny aircraft at a distance is not the easiest thing to do, make use of any hints you get over the radio from other pilots. A pilot reporting his position to a tower is also reporting to you. Your job is much easier when an air traffic controller tells you your traffic is "three miles at one o'clock." Once you have that particular traffic, don't forget the rest of the sky. If your traffic seems to be moving, you're not on a collision course, so continue your scan and watch it from time to time. If it doesn't appear to have motion, however, you need to watch it very carefully, and get out of the way, if necessary.

## Scan, Scan, Scan!

The most important part of your checklist, of course, is to keep looking where you're going and to watch for traffic. Make use of your scan constantly.

Basically, if you adhere to good airmanship, keep yourself and your plane in good condition, and develop an effective scan time-sharing system, you should have no trouble avoiding in-flight collisions. As you learn to use your eyes properly, you will benefit in other ways. Remember, despite their limitations, your eyes provide you with color, beauty, shape, motion and excitement. As you train them to spot minute targets in the sky, you'll also learn to see many other important "little" things you may now be missing, both on the ground and in the air. If you couple your eyes with your brain, you'll be around to enjoy these benefits of vision for a long time.

**4 - 5 seconds panel scan**

*Scan Pattern Time Allocation*

© Kendall Hunt Publishing Company

## Chapter Questions

1. The majority of in-flight collisions are the result of a _____ aircraft overtaking and hitting a _____ plane.
   A) faster, slower
   B) slower, faster
   C) faster, stationary

2. Approximately 77 percent of all midair collisions have occurred at or below _____feet (with 49 percent at or below _____ feet)
   A) 2,000 feet, 200 feet
   B) 3,000 feet, 500 feet
   C) 1,000 feet, 100 feet

3. In-flight collisions generally occur in the traffic pattern and primarily _____.
   A) Entering downwind
   B) on departure
   C) on final approach

4. Collisions occurring enroute generally are at or below 8,000 feet and within 25 miles of an airport.
   A) 1,000 feet, 50 miles
   B) 2,000 feet, 50 miles
   C) 8,000 feet, 25 miles

5. Explain at least three limitations of the eye when trying to avoid a midair collision.

# Bird Strike Mitigation

Bird strikes are a very prevalent hazard in aviation. In this chapter we will look at a prominent accident that was caused by bird strikes. We will explore where birds are most prevalent, and look into how we can best avoid birds and what to do if we have a bird strike.

On January 15, 2009, USAirways flight 1549, an Airbus A320, collided with birds while on initial climb after departure from New York's LaGuardia Airport. The flight encountered a flock of Canadian Geese at 2,800 feet, taking multiple bird strikes, including to both engines. The aircraft lost power to both engines and the crew declared an emergency, with the flight crew deciding to eventually ditch the aircraft in the Hudson River. This event, which could have been a catastrophic loss of life, ended up with all 155 people on the aircraft surviving with very few injuries (Figure 6.1).

If this much damage can be caused to an Airbus A320 by Canadian Geese, just think about the damage that it can inflict on a small general aviation aircraft. Studies show that a lot of damage can occur. A 12-pound Canada goose struck by a 150-knot aircraft generates the force of a 1,000-lb weight dropped from a height of 10 feet.

**Figure 6.1** USAirways Flight 1549 After Ditching in the Hudson River in New York
*Source:* NTSB.gov

Statistics on bird strikes:

In the United States alone, there were over 138,000 bird strikes to aircraft during the period of time from 1990 to 2013. This data is according to the FAA.

According to Richard Dolbeer, U.S. Department of Agriculture Wildlife Division: During the period of 1988–2005 . . .

**Aircraft lost to bird strikes worldwide:**
- Aircraft 144
- Fatalities 192

**Highest altitude reported bird strikes:**
- United States 32,000 Feet
- Africa 37,000 Feet

**Most prevalent time of year in U.S.:**
- Below 500' – July through October
- Above 500' – April and May and September through November

Some additional data from Richard Dolbeer, U.S. Department of Agriculture Wildlife Division. This data is from 1990–2004:

38,961 Reported Bird Strikes

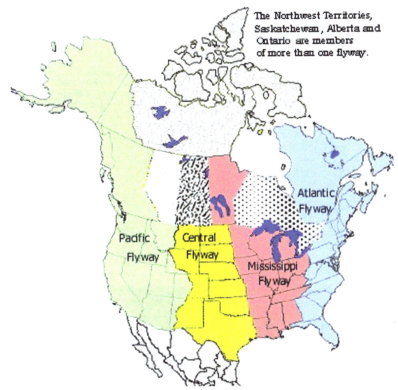

The Northwest Territories, Saskatchewan, Alberta and Ontario are members of more than one flyway.

Pacific Flyway

Central Flyway

Mississippi Flyway

Atlantic Flyway

**Figure 6.2** Overview of North American Flyways

*Source:* U.S. Fish & Wildlife Service

- 74% within 500 feet of ground.
- 19% from 501–3500 feet AGL
- 7% above 3500 Feet AGL

Strikes decrease 32% for every 1000 foot gain in altitude from 500' to 20,500 feet. Waterfowl comprised 53% of strikes above 3500 feet.

## 1.8 times more strikes at night vs. day.

Knowing where birds are more likely to be is a good way to be more cognizant of where the issue may lie. In North America there are four major flyways, or migration routes, for birds (Figures 6.2, 6.3).

Taking a look at two of these flyways, the Central and Mississippi Flyways, we find that:

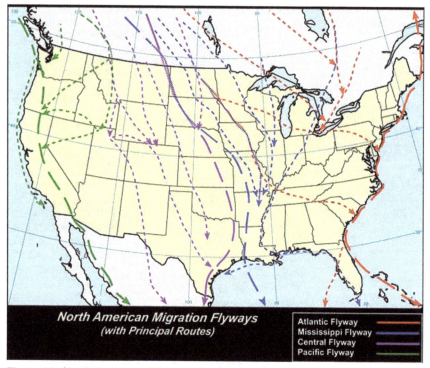

**Figure 6.3** North American Migration Flyways
*Source:* U.S. Fish & Wildlife Service

- No two species follow the same path from beginning to end.
- Generally a north-and-south movement with lanes of heavy concentration near rivers that are aligned north/south.
- Route primarily used by ducks, geese, blackbirds and sparrows.

Now that we have taken a look at the statistics on bird strikes, what can we do to reduce are chances of a bird strike, or even avoid them all together? Here are ten steps to Bird Strike Avoidance/Disaster:

- Plan your flight profile to climb from airports thru 500′ AGL as quickly and safely as possible when birds are present.
- Avoid low altitudes on cross country flight as much as feasible.
- Migratory strikes are most likely to occur in April and May and again in September through October.
- Even though it may change – report bird flock locations.

- Reduce speed at low altitudes – strikes (impact) forces increase as a function of the square of the aircraft velocity.
- Plan to climb. Birds almost invariably dive away, but there are always exceptions.
- If a collision seems likely, duck below the glareshield.
- If a collision occurs, fly the airplane first! Assess the damage. Then:
  - Land as soon as practical.
  - Land as soon as possible.
- Always have maintenance inspect the aircraft.
- File a safety report.

As number ten in this list is to file a safety report, where should that be accomplished? Always file the required safety report with your organization, but it is also best to fill out an FAA Bird/Wildlife Strike Form. This form can be found at this link: http://wildlife.faa.gov/strikenew.aspx

In conclusion – In North America, bird strike hazards are increasing. Because of outstanding wildlife conservation and environmental programs in North America, populations of many bird species have increased dramatically since the 1970s. Millions of acres have been set aside as wildlife refuges and strong environmental laws such as the Migratory Bird Treaty Act and the Federal Insecticide, Fungicide and Rodenticide Act have protected birds and other wildlife. As a result, species like non-migratory Canada Geese, which frequent urban areas such as golf courses, parks, and airports, have more than quadrupled in number since 1985. These increases have led to an increase in the number of birds in the vicinity of both large and small airports and greater opportunities for birds, especially larger birds, to hit aircraft.

## Chapter Questions

1. Bird strikes are more likely to happen:
   A) Less than 500'agl
   B) 500' to 3,500' aglAbove 3,500' agl
   C) None of the above
2. Which of the following is a North American flyway?
   A) Pacific
   B) Mississippi
   C) Central
   D) All of the above
3. There have been over 100,000 bird strikes in the U.S. alone since 1990:
   A) True
   B) False

# Hazardous Weather

## LEARNING OBJECTIVES

1. Understand the dangers of flying near thunderstorms.
2. Understand the effects of Low-Level Wind Shear and how to avoid Low-Level Wind Shear.
3. Understand the types and dangers of icing.
4. Understand how cold weather affects your altimeter and absolute altitude.

Benjamin Franklin aptly remarked: "Some people are weather-wise but most are otherwise." He wisely understood that weather affects all facets of life. Virtually all of our activities are affected by weather, but none is affected more than aviation. Aviation weather can be as uneventful as a clear and a million day or as challenging as descending through a solid deck of nimbostratus clouds with moderate rime icing and embedded thunderstorms. Aviators can narrow the uncertainty surrounding weather with a well-rounded understanding of weather processes. They can anticipate and avoid potential or existing hazardous weather conditions, and take advantage of favorable conditions such as tail winds, or clearing weather behind a cold front.

This chapter explains basic meteorological concepts and common weather problems. Aviation professionals can use this knowledge to ask weather forecasters critical questions such as the expected movement of reported severe thunderstorms or the length of time the visibility is to stay below 1/4 mile. What adjustments must be made if forecast target weather is marginal? At any rate, the better you understand weather, the safer your flight will be.

## Thunderstorms

For a thunderstorm to form, the air must have sufficient water vapor, an unstable lapse rate, and an initial lifting action to start the storm process. Some storms occur at random in unstable air, last for only an hour or two, and produce only moderate wind gusts and rainfall. These are known as airmass thunderstorms and are generally a result of surface heating. Steady-state thunderstorms are associated with weather systems. Fronts, converging winds, and troughs aloft force upward motion spawning these storms which often form into squall lines. In the mature stage, updrafts become stronger and last much longer than in air mass storms, hence the name steady state. (Figure 7.1) Knowledge of thunderstorms and the hazards associated with them is critical to the safety of flight.

### Hazards

Weather can pose serious hazards to flight and a thunderstorm packs just about every weather hazard known to aviation into one vicious bundle. These hazards occur individually or in combinations and most can be found in a squall line.

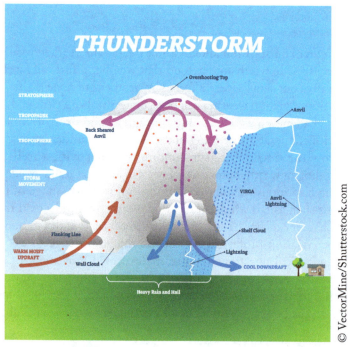

**Figure 7.1** Thunderstorm

## Squall Line

A squall line is a narrow band of active thunderstorms. Often it develops on or ahead of a cold front in moist, unstable air, but it may develop in unstable air far removed from any front. The line may be too long to detour easily and too wide and severe to penetrate. It often contains steady-state thunderstorms and presents the single most intense weather hazard to aircraft. It usually forms rapidly, generally reaching maximum intensity during the late afternoon and the first few hours of darkness.

## Tornadoes

The most violent thunderstorms draw air into their cloud bases with great vigor. If the incoming air has any initial rotating motion, it often forms an extremely concentrated vortex from the surface well into the cloud. Meteorologists have estimated that wind in such a vortex can exceed 200 knots with pressure inside the vortex quite low. The strong winds gather dust and debris and the low pressure generates a funnel-shaped cloud extending downward from the cumulonimbus base. If the cloud does not reach the surface, it is a funnel cloud; if it touches a land surface, it is a tornado. Tornadoes occur with both isolated and squall line thunderstorms. Reports for forecasts of tornadoes indicate that atmospheric conditions are favorable for violent turbulence. An aircraft entering a tornado vortex is almost certain to suffer structural damage. Since the vortex extends well into the cloud, any pilot inadvertently caught on instruments in a severe thunderstorm could encounter a hidden vortex. Families of tornadoes have been observed as appendages of the main cloud extending several miles outward from the area of lightning and precipitation. Thus, any cloud connected to a severe thunderstorm carries a threat of violence.

## Turbulence

Potentially hazardous turbulence is present in all thunderstorms, and a severe thunderstorm can destroy an aircraft. Strongest turbulence within the cloud occurs with shear between updrafts and downdrafts. Outside the cloud, shear turbulence has been encountered several thousand feet above and 20 miles laterally from a severe storm. A low-level turbulent area is the shear zone associated with the gust front.

Often, a "roll cloud" on the leading edge of a storm marks the top of the eddies in this shear and it signifies an extremely turbulent zone. Gust fronts often move far ahead (up to 15 miles) of associated precipitation. The gust front causes a rapid and sometimes drastic change in surface wind ahead of an approaching storm. Advisory Circular (AC) 00-50A, Low Level Wind

Shear, explains in detail the hazards associated with gust fronts. Figure 1 in the AC shows a schematic truss section of a thunderstorm with areas outside the cloud where turbulence may be encountered.

## Icing

Updrafts in a thunderstorm support abundant liquid water with relatively large droplet sizes. When carried above the freezing level, the water becomes supercooled. When temperature in the upward current cools to about -15 °C, much of the remaining water vapor sublimates as ice crystals. Above this level, at lower temperatures, the amount of supercooled water decreases. Supercooled water freezes on impact with an aircraft. Clear icing can occur at any altitude above the freezing level, but at high levels, icing from smaller droplets may be rime or mixed rime and clear ice. The abundance of large, supercooled water droplets makes clear icing very rapid between 0°C and -15°C and encounters can be frequent in a cluster of cells. Thunderstorm icing can be extremely hazardous. Thunderstorms are not the only area where pilots could encounter icing conditions. Pilots should be alert for icing anytime the temperature approaches 0°C and visible moisture is present.

## Hail

Hail competes with turbulence as the greatest thunderstorm hazard to aircraft. Supercooled drops above the freezing level begin to freeze. Once a drop has frozen, other drops latch on and freeze to it, so the hailstone grows – sometimes into a huge ice ball. Large hail occurs with severe thunderstorms with strong updrafts that have built to great heights. Eventually, the hailstones fall, possibly some distance from the storm core. Hail may be encountered in clear air several miles from thunderstorm clouds. As hailstones fall through air whose temperature is above 0°C, they begin to melt and precipitation may reach the ground as either hail or rain. Rain at the surface does not mean the absence of hail aloft. Possible hail should be anticipated with any thunderstorm, especially beneath the anvil of a large cumulonimbus. Hailstones larger than one half inch in diameter can significantly damage an aircraft in a few seconds.

## Ceiling and Visibility

Generally, visibility is near zero within a thunderstorm cloud. Ceiling and visibility also may be restricted in precipitation and dust between the cloud base and the ground. The restrictions create the same problem as all ceiling and visibility restrictions; but the hazards are multiplied when associated with the other thunderstorm hazards of turbulence, hail, and lightning.

## Effect on Altimeters

Pressure usually falls rapidly with the approach of a thunderstorm, rises sharply with the onset of the first gust and arrival of the cold downdraft and heavy rain showers, and then falls back to normal as the storm moves on. This cycle of pressure change may occur in 15 minutes. If the pilot does not receive a corrected altimeter setting, the altimeter may be more than 100 feet in error.

## Lightning

A lightning strike can puncture the skin of an aircraft and damage communications and electronic navigational equipment. Although lightning has been suspected of igniting fuel vapors and causing an explosion, serious accidents due to lightning strikes are rare. Nearby lightning can blind the pilot, rendering him or her momentarily unable to navigate either by instrument or by visual reference. Nearby lightning can also induce permanent errors in the magnetic compass. Lightning discharges, even distant ones, can disrupt radio communications on low and medium frequencies. Though lightning intensity and frequency have no simple relationship to other storm parameters, severe storms, as a rule, have a high frequency of lightning.

## Low-Level Wind Shear

Wind shear is a sudden, drastic change in wind speed and/or direction over a very small area. Wind shear can subject an aircraft to violent updrafts and downdrafts, as well as abrupt changes to the horizontal movement of the aircraft. While wind shear can occur at any altitude, low-level wind shear is especially hazardous due to the proximity of an aircraft to the ground. Directional wind changes of 180° and speed changes of 50 knots or more are associated with low-level wind shear. Low-level wind shear is commonly associated with passing frontal systems, thunderstorms, and temperature inversions with strong upper level winds (greater than 25 knots). Wind shear is dangerous to an aircraft for several reasons. The rapid changes in wind direction and velocity change the wind's relation to the aircraft disrupting the normal flight attitude and performance of the aircraft. During a wind shear situation, the effects can be subtle or very dramatic depending on wind speed and direction of change. For example, a tailwind that quickly changes to a headwind causes an increase in airspeed and performance. Conversely, when a headwind changes to a tailwind, the airspeed rapidly decreases and there is a corresponding decrease in performance. In either case, a pilot must be prepared to react immediately to the changes to maintain control of the aircraft. In general, the most severe type of low-level wind shear is associated with convective precipitation or rain from thunderstorms. One critical type of shear associated with convective precipitation is known as a microburst. A typical microburst occurs in a space of less than one mile horizontally and within 1,000 feet vertically. The lifespan of a microburst is about 15 minutes during which it can produce downdrafts of up to 6,000 feet per minute (fpm). It can also produce a hazardous wind direction change of 45 degrees or more, in a matter of seconds. When encountered close to the ground, these excessive downdrafts and rapid changes in wind direction can produce a situation in which it is difficult to control the aircraft. (Figure 7.2)

During an inadvertent takeoff into a microburst, the plane first experiences a performance increasing headwind (1), followed by performance-decreasing downdrafts (2). Then, the wind rapidly shears to a tailwind (3), and can result in terrain impact or flight dangerously close to the ground (4). Microbursts are often difficult to detect because they occur in relatively confined areas. In an effort to warn pilots of low-level wind shear, alert systems have been installed at several airports around the country. A series of anemometers, placed around the airport, form a net to detect changes in wind speeds. When wind speeds differ by more than 15 knots, a warning for wind shear is given to pilots. This system is known as the low-level wind shear alert system (LLWAS). It is important to remember that wind shear can affect any flight and any pilot at any altitude. While wind shear may be reported, it often remains undetected and is a silent danger to aviation. Always be alert to the possibility of wind shear, especially when flying in and around thunderstorms and frontal systems.

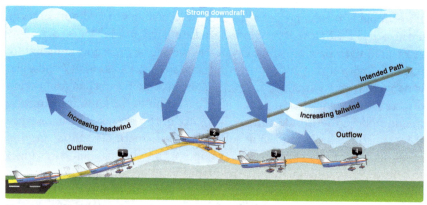

**Figure 7.2** Windshear
*Source:* FAA.gov

# Icing

## Aircraft Icing

Aircraft icing is a major weather hazard to aviation. Many aircraft accidents and incidents have been attributed to aircraft icing. In fact, many icing-related mishaps have occurred when the aircraft was not deiced before takeoff. Most of the time, ground deicing and anti-icing procedures will adequately handle icing formation, but there are times when you may be caught unaware of dangerous ice buildup. This chapter will help you understand icing formation processes and what you can do if suddenly caught in an icing situation.

Icing formation on either fixed or rotary wing aircraft disrupts the flow of air over the airfoils increasing weight and stalling speed. Test data indicates that icing reduces wing lift by up to 30 percent and increases drag by 40 percent. The accumulation of ice on exterior movable surfaces also affects the control of the aircraft. If ice begins forming on a propeller's blades, the propeller's efficiency decreases and requires further power to maintain flight. Another significant hazard comes from ice accumulation on rotors and propellers resulting in disastrous vibrations. Ice can also form in an engine's intake, robbing the engine of air needed to support combustion, or ice may break off and may be ingested into the engine, causing foreign object damage (FOD). Other icing effects include loss of proper operation of control surfaces, brakes, or landing gear; reduction or loss of aircrew's outside vision, false flight instrument indications, and loss of radio communication.

## Groups of Icing

Aircraft icing is classified into two main groups: structural and induction. We will discuss these icing groups in detail to include conditions contributing to ice formation, icing intensities, icing types, and where icing is most likely found.

## Structural Icing

Two conditions must be met for structural ice to form on an aircraft. First, the air and aircraft's surface temperatures must be at or below freezing. (Instances of freezing precipitation are exempt from this rule.) Second, supercooled, visible water droplets (liquid water droplets at subfreezing temperatures) must be present or high humidity must exist.

Wind tunnel experiments reveal that saturated air flowing over a stationary object may form ice on the object when the air temperature is as high as 4°C. The object's temperature cools by evaporation and pressure changes in the moving air currents. Conversely, friction and water droplet impacts heat the object. When an aircraft travels at about 400 knots true airspeed, the cooling and heating effects tend to balance. Structural ice may form when the free-air temperature is 0°C or colder. Icing is seldom encountered below -40°C.

Clouds are the most common forms of visible liquid water. Water droplets in the free air, unlike bulk water, do not freeze at 0°C. Instead, their freezing temperature varies from -10 to -40°C. The smaller the droplets, the lower the freezing point. As a general rule, serious icing is rare in clouds with temperatures below -20°C since these clouds are almost completely composed of ice crystals. However, be aware that icing is possible in any cloud when the temperature is 0°C or below. In addition, frost may form on an aircraft in clear, humid air if the aircraft skin temperature is below freezing.

Freezing rain and drizzle, sometimes found in the clear air below a cloud deck, are other forms of visible liquid moisture causing icing. Freezing precipitation is the most dangerous of all icing conditions. It can build hazardous amounts of ice in a few minutes and is extremely difficult to remove.

## Types of Icing

Aircraft structural icing consists of three basic types: clear, rime and mixed. Frost is another form of icing, but is not forecasted as a type of icing. Icing types that form will depend primarily upon the water droplet size and temperature.

### Clear Ice

Clear ice is a glossy ice identical to the glaze forming on trees and other objects as freezing rain strikes the Earth. Clear ice is the most serious of the various forms of ice because it adheres so firmly to the aircraft. Conditions most favorable for clear ice formation are high water content, large droplet size, and temperatures slightly below freezing. Clear ice normally forms when temperatures are between 0o and -16°C, and is most frequently forecasted in cumuliform clouds between 0o and -08°C and during freezing precipitation. Clear icing can also be encountered in cumulonimbus clouds in temperatures as low as -25°C.

Clear ice can be smooth or rough. It is smooth when deposited from large, supercooled cloud droplets or raindrops that spread, adhere to the surface of the aircraft and slowly freeze. If mixed with snow, ice pellets or small hail, it is rough, irregular, and whitish (Figure 7.3). The deposit then becomes very blunt-nosed with rough bulges building out against the airflow. Clear ice is hard, heavy, and tenacious. Its removal by deicing equipment is especially difficult.

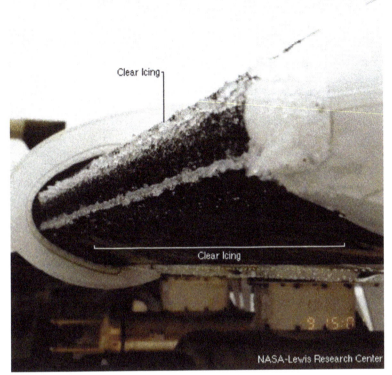

Clear Icing

Clear Icing

NASA-Lewis Research Center

**Figure 7.3** Clear Ice Can be Smooth or Rough
*Source:* NASA.gov

## Rime Ice

Rime ice is a milky, opaque, and granular deposit with a rough surface (Figure 7.4). It forms by the instantaneous freezing of small, supercooled water droplets as they strike the aircraft. This instantaneous freezing traps a large amount of air, giving the ice its opaqueness and making it very brittle. Rime ice is most frequently encountered in stratiform clouds but also occurs in cumulus clouds. Rime ice may form in stratiform clouds from 0° to -30°C, but most frequently occurs within stratus clouds between -08° and -10°C. It may also accumulate when temperatures in cumuliform clouds are between 0° and -20°C but can be expected in thunderstorms as cold as -40°C. Rime ice is lighter in weight than clear ice and its weight is of little significance. Rime ice is brittle and more easily removed than clear ice.

Mixed icing forms when water drops vary in size or when liquid drops are intermingled with snow or ice particles. It can form rapidly. Ice particles become embedded in clear ice, building a very rough accumulation sometimes in a mushroom shape on leading edges. Mixed icing is generally forecasted

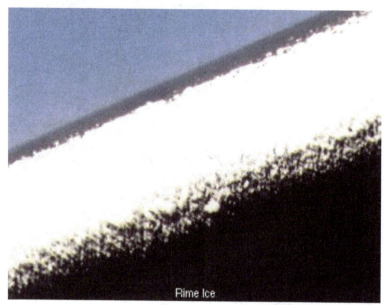

**Figure 7.4** Rime Ice is Milky, Opaque, and Granular
*Source:* NASA.gov close up

at temperatures between -9° and -15°C, and is commonly encountered between -10° and -15°C.

## Frost

Frost is deposited as a thin layer of crystalline ice (Figure 7.5). It forms on the exposed surfaces of parked aircraft when the temperature of the exposed surface is below freezing (although the air temperature may be above freezing). The deposit forms during night radiational cooling in the same way the formation of frost found on the ground. Frost may also form on aircraft in flight when a cold aircraft moves from a zone of subzero temperatures to a warmer, moist layer. Contact with the cold aircraft suddenly chills the air to below freezing temperatures and deposition (formation of ice crystals directly from water vapor) occurs. Frost can cover the windshield or canopy and completely restrict outside vision. It also affects the aircraft's lift to drag ratio and can be a hazard during takeoff. Remove all frost from the aircraft prior to departure.

## Icing Amounts

The amount of ice an aircraft accumulates depends considerably on the characteristics of that particular aircraft. Therefore, general intensity classifications for reporting icing are given in the Meteorological Information section of the Flight Information Handbook (FIH) and are described below.

**Figure 7.5** Frost on Exposed Surfaces of Parked Aircraft

### Trace Ice

Trace – Ice becomes perceptible. The rate of accumulation is slightly greater than rate of sublimation. It is not hazardous unless encountered for an extended period of time (over one hour) even though de-icing/anti-icing equipment is not used.

### Light Ice

Light – The rate of accumulation may create a problem if flight is prolonged in this environment (over one hour). Occasional use of de-icing/anti-icing equipment removes/prevents accumulation. It does not present a problem if the de-icing/anti-icing equipment is used.

### Moderate Ice

Moderate – The rate of accumulation is such that even short encounters become potentially hazardous and use of de-icing/anti-icing equipment or diversion is necessary.

### Severe Ice

Severe – The rate of accumulation is such that de-icing/anti-icing equipment fails to reduce or control the hazard. Immediate diversion is necessary.

### Icing Dangers

The relatively thick wings, canopies, and other features of conventional aircraft have a smaller collection potential than those of the trimmer and faster turbojet aircraft. However, the actual hazard of icing for conventional aircraft tends to be greater than for jets because of less aerodynamic heating at lower airspeed. Conventional aircraft are subjected to icing conditions over longer periods and operate at altitudes more conducive to icing. Ice accumulations on wing and tail surfaces disrupt the air flow around these airfoils. This results in a loss of lift, an increase in drag, and causes higher than normal stall speeds (Figure 7.6). The weight of the ice deposit presents less danger, but may become important when too much lift and thrust are lost. Experiments have shown that a ½ inch ice deposit on the leading edge of airfoils on some aircraft reduce their lift by as much as 50 percent and increases drag on the aircraft by the same amount, which greatly increases the stall speed. The serious consequences of these effects are obvious. Remember that ½ inch or more of ice can accumulate in a minute or two.

Ice accumulation on the propeller hub and blades reduces the propeller's efficiency, which reduces thrust. Increased power settings consume more fuel and may fail to produce sufficient thrust to maintain altitude. An even greater hazard is the vibration of the propeller, caused by the uneven distribution of ice on the blades. A propeller is very delicately balanced, and even a small amount of ice creates an imbalance. The resulting vibration places dangerous stress on the engine mounts as well as the propeller itself. Propellers with low

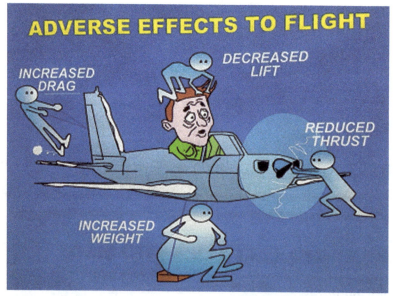

**Figure 7.6** Effects of Icing are Cumulative Causing Stall Speed to Increase

*Source:* The National Weather Service (NWS)

RPM are more susceptible to icing than those with high RPM. Ice usually forms faster on the propeller's hub because the blade's differential velocity causes a temperature increase from the hub to the propeller tip.

Icing of the pitot tube (Figure 7.7) and static pressure ports is dangerous because it causes inaccurate indications on the altimeter, airspeed, and VSI. When icing is observed on the aircraft, remember that the pitot tubes accumulate ice as fast as or faster than other areas of the aircraft.

The principal danger of ice accumulating on the aircraft's radio antenna is the probable loss of radio communication. Antennas are usually one of the first items on an aircraft to collect ice. Other parts of the aircraft will also begin to accumulate ice if the antennas start icing up. Ultimately, aircrews lose their ability to request altitude or course changes to get out of the icing zone.

Ice or frost formation on an aircraft's windshield is most hazardous during takeoffs and landings. Small frost particles on the windshield prior to takeoff may act as sublimation nuclei during takeoff and reduce visibility to near zero. On approach, windshield icing may prevent visual contact with the runway. In large helicopters, windshield icing is a good indication that main rotor head and rotor blade icing is well underway. Reciprocating engines experience icing on air scoops, scoop inlets (ducts), carburetor inlet screens and other induction system protuberances. All surfaces of the engine exposed to water droplets may collect ice.

**Figure 7.7** Pitot Tube Icing
*Source:* Nasa.gov

## Helicopter Icing

Icing on rotary wing aircraft is related to those involving wings and propellers. Rotor icing is slightly different from propeller icing due to the rotors' lower rotational speed. Ice accumulation on rotor blades differs from the fixed wings of conventional aircraft due to the smaller scale of the helicopter wing, the variation of airspeed with rotor blade span, the cyclic pitch change, and the cyclic variation of airspeed at any given point on the blade in forward flight. Ice formation on the helicopter main rotor system or anti-torque rotor system may produce serious vibration, loss of efficiency or control, and can significantly deteriorate the available RPM to a level where safe landing cannot be assured. Although the slow forward speed of the helicopter reduces ice build-up on the fuselage, the rotational speed of main and tail rotor blades produces a rapid growth rate on certain surface areas. Ice accumulation on the swash plates, push-pull rods, bell cranks, hinges, scissors assemblies, and other mechanisms of the main rotor head assembly interferes with collective and cyclic inputs.

Several factors tend to reduce ice accretion on the main rotor blades, such as the centrifugal force of rotation, blade flexing during rotation, the slow rotational speed of the blades near the rotor head, and the fast rotational speed near the blade tips. However, in a hover, a 3/16 inch coating of ice is sufficient to prevent some helicopters from maintaining flight. A critical icing hazard can, therefore, form rapidly on the center two-thirds of the main rotor blades. The uneven accretion or asymmetrical shedding of ice produces severe rotor vibration. Ice accumulation on either the antitorque rotor head assembly or blades produces the same hazards as those associated with the main rotor. The centrifugal force of rotation and the blade angle of incidence relative to the clouds help to reduce ice build-up on the tail rotor blades, but the shedding of ice from the blades may result in structural damage or FOD to the fuselage, rotors or engines, and injury to ground personnel. This particular hazard appears to be more threatening to large, tandem rotor aircraft.

Ice accumulation on the engine and transmission air intake screens is more rapid than on the rotor systems. This results in inadequate cooling of the engine and transmission. On some helicopters, a loss of manifold pressure concurrently with air intake screen icing may force an immediate landing. Freezing water passing through the screens also coats control cables and may produce limited throttle movement and other control problems.

## Engine Icing

In addition to the hazards created by structural icing, aircraft are frequently subjected to engine icing. The affected components supply the engine with the proper fuel and air mixture for efficient combustion. Induction icing occurs under a wide range of weather conditions and is most common in the air induction system but may also be found in the fuel system. Carburetor icing in carburetor equipped piston engines is actually a combination of the two.

## Carburetor Icing

Carburetor icing is treacherous. It frequently causes complete engine failure. It may form under conditions in which structural ice could not possibly form. If the air drawn into the carburetor has a high relative humidity, ice can form inside the carburetor in cloudless skies with temperatures as high as 22°C (72°F). It sometimes forms with outside air temperatures as low as -10°C (14°F). Carburetor ice forms during fuel vaporization, combined with the air expanding as it passes through the carburetor. Temperature drop in the carburetor can be as much as 40°C but is usually 20°C or less. With enough available moisture, ice will form in the carburetor passages (Figure 7.8) if the temperature inside the carburetor cools down to 0°C or below. Ice may form at the discharge nozzle, in the Venturi, on or around the butterfly valve, or in the passages from the carburetor to the engine.

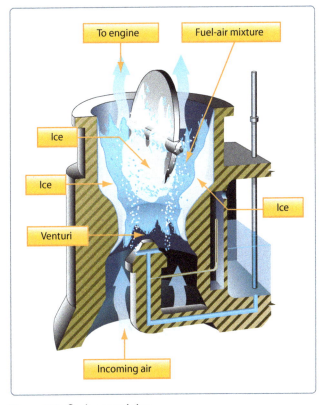

**Figure 7.8** Carburetor Icing
*Source:* FAA.gov

The carburetor heater is an anti-icing device which heats the air before it reaches the carburetor, melting any ice or snow entering the intake, and keeping the mixture above the freezing point. The heater usually prevents icing, but it cannot always clear out ice already formed. Since carburetor heating adversely affects aircraft performance, use it only as specified in your flight manual. The fuel absorbs considerable amounts of water when the humidity is high. Occasionally, enough water is absorbed to create icing in the fuel system when the fuel temperature is at or below 0°C.

## Induction Icing

Ice forms in the induction system when atmospheric conditions are favorable for structural icing (visible liquid moisture and freezing temperatures). Induction icing can form in clear air with a high relative humidity (small temperature/dew point spread) and temperatures anywhere from 22°C (72°F) to -10°C (14° F).

In flights through clouds containing supercooled water droplets, air intake duct icing is similar to wing icing. However, duct icing may occur with clear skies and above freezing temperatures. While taxiing, and on departure, reduced pressures exist in the intake system (Figure 7.9). This lowers temperatures to the point where condensation or sublimation takes place, resulting in ice formation which decreases the radius of the duct opening and limits air intake.

The temperature change varies considerably with different types of engines. Therefore, if the air temperature is 10°C or less (especially near the

© Jaromir Chalabala/Shutterstock.com

**Figure 7.9** Jet Engine Induction Icing

freezing point) and the relative humidity is high, the possibility of induction icing definitely exists.

## Inlet Guide Vane Icing

Icing occurs when supercooled water droplets in the atmosphere strike the guide vanes and freeze. As ice build-up increases, air flow to the engine decreases, which results in a loss of thrust and eventual engine flameout. Also, ingestion of ice shed ahead of the compressor inlet may cause severe engine damage.

## Weather Conditions for Icing

Potential icing zones in the atmosphere are mainly a function of temperature and cloud structure. These factors vary with altitude, location, weather pattern, season, and terrain.

Generally, aircraft icing is limited to the atmospheric layer lying between 0°C and -20°C. However, icing has been reported at temperatures colder than -40°C in the upper parts of cumulonimbus and other clouds. Table 7.1 shows the types of icing in cumuliform clouds associated with variable temperatures:

Icing in middle and low level stratiform clouds is confined, on the average, to a layer between 3,000 and 4,000 feet thick. Icing intensity generally ranges from a trace to light, with the maximum values occurring in the cloud's upper portions. Both rime and mixed are found in stratiform clouds. The main hazard lies in the great horizontal extent of these cloud decks. High-level stratiform clouds are composed mostly of ice crystals and give little icing. The icing zone in cumuliform clouds is smaller horizontally but greater vertically than in stratiform clouds. Icing is more variable in cumuliform clouds because many of the factors conducive to icing depend on the particular cloud's stage of development. Icing intensities may range from a trace in a small cumulus to severe in a large towering cumulus or cumulonimbus. Although icing occurs at all levels above the freezing level in a building cumulus, it is most intense in the upper half of the cloud. Icing in a cumuliform cloud is usually clear or mixed with rime in the upper levels. Aircraft icing rarely occurs in cirrus clouds although some do contain a small

**Table 7.1** Temperature Ranges

| 0°C to -10°C | clear |
|---|---|
| -10°C to -15°C | missed, clear, and rime |
| -15°C to 20°C and colder | rime. |

portion of water droplets. However, light icing has been reported in the dense, cirrus anvils of cumulonimbus, where updrafts maintain considerable amounts of water at rather low temperatures.

Of all icing conditions reported, 85 percent occur in the vicinity of fronts. This icing may be in relatively warm air above the frontal surface or in the cold air beneath (Figures 7.10 and 7.11).

For significant icing to occur above the frontal surface, the warm air must be lifted and cooled to saturation at temperatures below freezing, making it contain supercooled water. If the warm air is unstable, icing may be sporadic; if it is stable, icing may be continuous over an extended area. Icing may form in this manner over either a warm or a shallow cold frontal surface. A line of showers or thunderstorms along a surface cold front may produce icing, but only in a comparatively narrow band along the front (Figure 7.12).

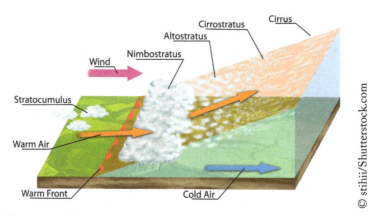

**Figure 7.10** Warm Front Icing

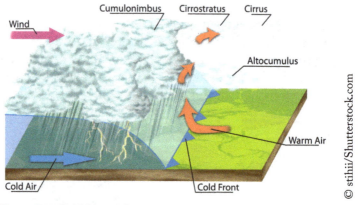

**Figure 7.11** Cold Front Icing

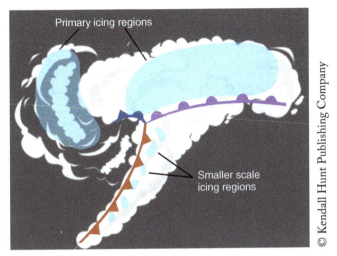

Figure 7.12 Primary Icing Regions

Icing below a frontal surface outside of clouds occurs most often in freezing rain or drizzle. Precipitation forms in the relatively warm air above the frontal surface at temperatures above freezing. It falls into the subfreezing air below the front, supercools and freezes on impact with the aircraft. Freezing drizzle and rain occur with both warm and shallow cold fronts. Icing in freezing precipitation is especially hazardous since it often extends horizontally over a broad area and downward to the surface.

## Affects of Temperature on Pressure

The advent of aviation early in the last century brought about a search for an accurate method of measuring the altitude at which an aircraft was flying. Barometric pressure was ideal for several reasons, chiefly the fact that pressure change with altitude is approximately 10,000 times greater than that found in equivalent horizontal distances. The rate of change in vertical heights in the lower atmosphere is about 1 inch for every 1,000 feet of altitude. An aircraft altimeter (Figure 7.13) is essentially an aneroid barometer calibrated to indicate altitude in feet instead of pressure. This altitude is independent of the terrain below. An altimeter reads accurately only in a standard atmosphere and when properly adjusted altimeter settings are used. Remember, an altimeter is only a pressure-measuring device. It indicates 10,000 feet with 29.92 set in the Kollsman window, and the pressure is 697 millibars, whether or not the altitude is actually 10,000 feet.

The effect is important since in the lowest 15,000 feet of the atmosphere a 2.3°C deviation of the mean temperature from the standard

**Figure 7.13** Aneroid Altimeter

© Alex Vernon/Shutterstock.com

temperature of 2.8°C will cause about a 1% error in the altimeter reading. For example, if an aircraft with a correct altimeter setting is flying at an indicated altitude of 10,000 feet but the air below flight level is 11°C warmer than the standard atmosphere temperature of 2.8°C, the altimeter will read about 4 percent too low. The aircraft will be flying at a true altitude of 10,400 feet (400 feet higher than indicated). Carefully study Figures 7.14 and 7.15 to visually see the dangerous effects of temperature on aircraft indicated and true altitudes.

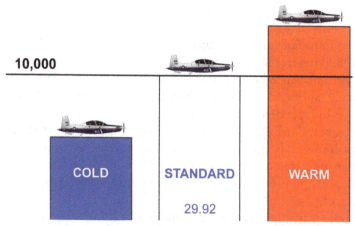

**Figure 7.14** Effects of Temperature on True and Indicated Altitude
*Source:* FAA.gov

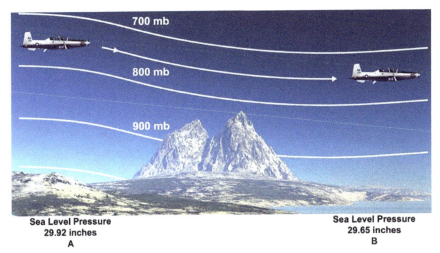

**Figure 7.15** Altitude Error Due to Nonstandard Temperatures Aloft

*Source:* FAA.gov

## Chapter Questions

1. Identify 5 hazards associated with thunderstorms
2. Explain trace icing
3. Explain moderate icing
4. Explain severe icing
5. Explain the weather conditions contusive to low level wind shear
6. Explain the effects on your absolute altitude when flying in extreme cold temperatures.

# The Human Factors Analysis and Classification System (HFACS) and the "Swiss Cheese Model of Accident Causation"

# 8

## LEARNING OBJECTIVES

1. Know the definition of the following terms: unsafe acts, preconditions to unsafe acts, errors, routine violations, exceptional violations, and unsafe supervision.
2. Understand the difference between active failures and latent failures.
3. Explain how an organization can be responsible for an accident.
4. Know the percentage of accidents caused by human error.

## Introduction

Sadly, the annals of aviation history are littered with accidents and tragic losses. Since the late 1950s, however, the drive to reduce the accident rate has yielded unprecedented levels of safety to a point where it is now safer to fly in a commercial airliner than to drive a car or even walk across a busy New York city street. Still, while the aviation accident rate has declined tremendously since the first flights over a century ago, the cost of aviation accidents in both lives and dollars has steadily risen. As a result, the effort to reduce the accident rate still further has taken on new meaning within both military and civilian aviation.

Why do aircraft crash? The answer is complicated. In the early years of aviation, it could reasonably be said that, more often than not, the aircraft killed the pilot. That is, aircraft are intrinsically unforgiving and unsafe. However, the modern era of aviation has witnessed an ironic reversal of sorts. It now appears that the aircrew themselves are deadlier than the aircraft they fly. In fact, estimates in the literature indicate that between 70 and 80 percent of aviation accidents can be attributed, at least in part, to human error.

So, what really constitutes that 70–80% of human error? Some would have us believe that human error and "pilot" error are synonymous. Yet, simply writing off aviation accidents merely to pilot error is an overly simplistic, if not naive, approach to accident causation. After all, it is well established that accidents cannot be attributed to a single cause, or in most instances, even a single individual. In fact, even the identification of a "primary" cause is fraught with problems. Rather, aviation accidents are the result of a number of causes, only the last of which are the unsafe acts of the aircrew.

The challenge for safety analysts and accident investigators alike is how best to identify and mitigate the causal sequence of events, in particular that 70–80% associated with human error. Additionally, human error does not only refer to errors made by the pilot. We also need to analyze human errors made by aircraft maintenance workers, flight attendants, air traffic controllers, weather forecasters, airport managers, dispatchers, etc.

### Reason's "Swiss Cheese" Model of Human Error

One particularly appealing approach to the genesis of human error is the one proposed by James Reason in 1990. Generally referred to as the "Swiss Cheese" model of human error. Reason's original work involved operators of a nuclear power plant. This model has been used to analyze human errors within the aviation community. Reason describes four levels of human failure, each influencing the next (Figure 8.1).

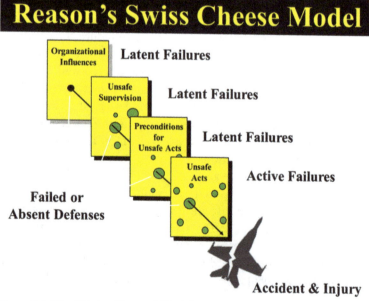

**Figure 8.1** The "Swiss Cheese" Model of Human Error Causation (Adapted from Reason, 1990)

Working backwards in time from the accident, the first level depicts those *Unsafe Acts* of Operators that ultimately led to the accident. Unsafe Acts reflect the active failures. More commonly referred to in aviation as aircrew/pilot error, these active failures has been the traditional focus of most accident investigations. Represented as "holes" in the cheese, these active failures are typically the last unsafe acts committed by aircrew.

However, what makes the "Swiss cheese" model particularly useful in accident investigation is that it forces investigators to address latent failures within the causal sequence of events as well. As their name suggests, latent failures, unlike their active counterparts, may lie dormant or undetected for hours, days, weeks, or even longer, until one day they adversely affect the unsuspecting aircrew. Consequently, investigators with even the best intentions often overlook them.

Within this concept of latent failures, Reason described three more levels of human failure. The first involves the condition of the aircrew as it affects performance. Referred to as *Preconditions for Unsafe Acts*, this level involves conditions such as mental fatigue and poor communication and coordination practices, often referred to as crew resource management (CRM). Not surprising, if fatigued aircrew fail to communicate and coordinate their activities with others in the cockpit or individuals external to the aircraft (e.g., air traffic control, maintenance, etc.), poor decisions are made and errors often result.

*But exactly why did communication and coordination break down in the first place?* This is perhaps where Reason's work departed from more traditional approaches to human error. In many instances, the breakdown in good CRM practices can be traced back to instances of *Unsafe Supervision*, the third level of human failure. If, for example, two inexperienced (and perhaps even below average pilots) are paired with each other and sent on a flight into known adverse weather at night, is anyone really surprised by a tragic outcome? To make matters worse, if this questionable manning practice is coupled with the lack of quality CRM training, the potential for miscommunication and ultimately, aircrew errors, is magnified. In a sense then, the crew was "set up" for failure as crew coordination and ultimately performance would be compromised. This is not to lessen the role played by the aircrew, only that intervention and mitigation strategies might lie higher within the system.

Reason's model didn't stop at the supervisory level either; the organization itself can impact performance at all levels. For instance, in times of fiscal austerity, funding is often cut, and as a result, training and flight time are curtailed. Consequently, supervisors are often left with no alternative but to task "non- proficient" aviators with complex tasks. Not surprisingly then, in the absence of good CRM training, communication and coordination failures will begin to appear as will a myriad of other preconditions, all of which will affect performance and elicit aircrew errors. Therefore, it makes sense that, if the accident rate is going to be reduced beyond current levels, investigators and analysts alike must examine the accident sequence in its entirety and expand it beyond the cockpit. Ultimately, causal factors at all levels within the organization must be addressed if any accident investigation and prevention system is going to succeed. In many ways, Reason's "Swiss cheese" model of accident causation has revolutionized common views of accident causation.

## Unsafe Acts

The unsafe acts of aircrew can be loosely classified into two categories: errors and violations. In general, errors represent the mental or physical activities of individuals that fail to achieve their intended outcome. Not surprising, given the fact that human beings by their very nature make errors, these unsafe acts dominate most accident databases. Violations, on the other hand, refer to the willful disregard for the rules and regulations that govern the safety of flight.

Still, distinguishing between errors and violations does not provide the level of granularity required of most accident investigations. Therefore, the categories of errors and violations were expanded here (Figure 8.2), as elsewhere, to include three basic error types (skill-based, decision, and perceptual) and two forms of violations (routine and exceptional). Table 8.1 identifies a few examples of unsafe acts.

### Violations

By definition, errors occur within the rules and regulations espoused by an organization; typically dominating most accident databases. In contrast, violations represent a willful disregard for the rules and regulations that govern safe flight and, fortunately, occur much less frequently since they often involve fatalities.

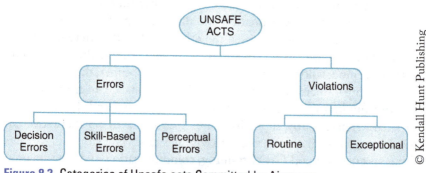

**Figure 8.2  Categories of Unsafe acts Committed by Aircrews**

© Kendall Hunt Publishing Company

**Table 8.1** Selected Examples of Unsafe Acts of Pilot Operators (Note: Not a Complete Listing)

| ERRORS | VIOLATIONS |
|---|---|
| **Skill based Errors** | ■ Failed to adhere to brief |
| ■ Breakdown in visual scan | ■ Failed to use the radar altimeter |
| ■ Failed to prioritize attention | ■ Flew an unauthorized approach |
| ■ Inadvertent use of flight controls | ■ Violated training rules |
| ■ Omitted step in procedure | ■ Flew an overaggressive maneuver |
| ■ Omitted checklist item | ■ Failed to properly prepare for the flight |
| ■ Poor technique | ■ Briefed unauthorized flight |
| ■ Over controlled the aircraft Decision | ■ Not current/qualified for the mission |
|  | ■ Intentionally exceeded the limits of the aircraft |
| **Errors** | ■ Continued low-altitude flight in VMC |
| ■ Improper procedure | ■ Unauthorized low-altitude canyon running |
| ■ Misdiagnosed emergency |  |
| ■ Wrong response to emergency |  |
| ■ Exceeded ability |  |
| ■ Inappropriate maneuver |  |
| ■ Poor decision Perceptual |  |
|  |  |
| **Errors (due to)** |  |
| ■ Misjudged distance/altitude/airspeed |  |
| ■ Spatial disorientation |  |
| ■ Visual illusion |  |

While there are many ways to distinguish between types of violations, two distinct forms have been identified, based on their etiology, that will help the safety professional when identifying accident causal factors. The first, routine violations, tend to be habitual by nature and often tolerated by governing authority. Consider, for example, the individual who drives consistently 5–10 mph faster than allowed by law or someone who routinely flies in marginal weather when authorized for visual meteorological conditions only. While both are certainly against the governing regulations, many others do the same thing. Furthermore, individuals who drive 64 mph in a 55-mph zone, almost always drive 64 in a 55 mph zone. That is, they "routinely" violate the speed limit. The same can typically be said of the pilot who routinely flies into marginal weather.

What makes matters worse, these violations (commonly referred to as "bending" the rules) are often tolerated and, in effect, known AND sanctioned by supervisory authority (i.e., you're not likely to get a traffic citation until you exceed the posted speed limit by more than 10 mph). If, however, the local authorities started handing out traffic citations for exceeding the speed limit on the highway by 9 mph or less (as is often done on military installations), then it is less likely that individuals would violate the rules.

Therefore, by definition, if a routine violation is identified, one must look further up the supervisory chain to identify those individuals in authority who are not enforcing the rules.

On the other hand, unlike routine violations, exceptional violations appear as isolated departures from authority, not necessarily indicative of individual's typical behavior pattern nor condoned by management. For example, an isolated instance of driving 105 mph in a 55-mph zone is considered an exceptional violation. Likewise, flying under a bridge or engaging in other prohibited maneuvers, like low-level canyon running, would constitute an exceptional violation. However, it is important to note that, while most exceptional violations are appalling, they are not considered "exceptional" because of their extreme nature. Rather, they are considered exceptional because they are neither typical of the individual nor condoned by authority. Still, what makes exceptional violations particularly difficult for any organization to deal with is that they are not indicative of an individual's behavioral repertoire and, as such, are particularly difficult to predict. In fact, when individuals are confronted with evidence of their dreadful behavior and asked to explain it, they are often left with little explanation. Indeed, those individuals who survived such excursions from the norm clearly knew that, if caught, dire consequences would follow. Still, defying all logic, many otherwise model citizens have been down this potentially tragic road.

## Preconditions for Unsafe Acts

Arguably, the unsafe acts of pilots can be directly linked to nearly 70–80% of all aviation accidents. However, simply focusing on unsafe acts is like focusing on a fever without understanding the underlying disease causing it. Thus, investigators must dig deeper into why the unsafe acts took place. As a first step, two major subdivisions of unsafe aircrew conditions were developed: substandard conditions of operators and the substandard practices they commit (Figure 8.3). Table 8.2 identifies a few examples of preconditions for unsafe acts.

### Inadequate Supervision

The role of any supervisor is to provide the opportunity to succeed. To do this, the supervisor, no matter at what level of operation, must provide guidance, training opportunities, leadership, and motivation, as well as the proper role model to be emulated. Unfortunately, this is not always the case.

For example, it is not difficult to conceive of a situation where adequate crew resource management training was either not provided, or the opportunity to attend such training was not afforded to a particular aircrew member. Conceivably, aircrew coordination skills would be compromised and if the

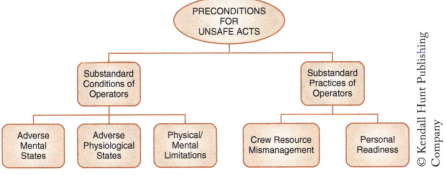

**Figure 8.3** Categories of Preconditions of Unsafe Acts

**Table 8.2** Selected Examples of Preconditions of Unsafe Acts (Note: This is not a Complete Listing)

| Substandard Conditions of Operators | Substandard Practice of Operators |
|---|---|
| **Adverse Mental States** | **Crew Resource Management** |
| ■ Channelized attention | ■ Failed to back-up |
| ■ Complacency | ■ Failed to communicate/coordinate |
| ■ Distraction | ■ Failed to conduct adequate brief |
| ■ Mental Fatigue | ■ Failed to use all available resources |
| ■ Get-home-itis | ■ Failure of leadership |
| ■ Haste | ■ Misinterpretation of traffic calls |
| ■ Loss of situational awareness | |
| ■ Misplaced motivation | **Personal Readiness** |
| ■ Task saturation Adverse | ■ Excessive physical training |
| | ■ Self-medicating |
| **Physiological States** | ■ Violation of crew rest requirement |
| ■ Impaired physiological state | ■ Violation of bottle-to-throttle requirement |
| ■ Medical illness | |
| ■ Physiological incapacitation | |
| ■ Physical fatigue | |
| | |
| **Physical/Mental Limitation** | |
| ■ Insufficient reaction time | |
| ■ Visual limitation | |
| ■ Incompatible intelligence/aptitude | |
| ■ Incompatible physical capability | |

aircraft were put into an adverse situation (an emergency for instance), the risk of an error being committed would be exacerbated and the potential for an accident would increase markedly.

In a similar vein, sound professional guidance and oversight is an essential ingredient of any successful organization. While empowering individuals to

**Figure 8.4** Categories of Unsafe Supervision/Inadequate Supervision

make decisions and function independently is certainly essential, this does not divorce the supervisor from accountability. The lack of guidance and oversight has proven to be the breeding ground for many of the violations that have crept into the cockpit. As such, any thorough investigation of accident causal factors must consider the role supervision plays (i.e., whether the supervision was inappropriate or did not occur at all) in the genesis of human error (Table 8.3).

### Inadequate Supervision

- Failed to provide guidance
- Failed to provide operational doctrine
- Failed to provide oversight

**Table 8.3** Selected examples of Unsafe Supervision (Note: This is not a Complete Listing)

| Inadequate Supervision | Failed to Correct a Known Problem |
|---|---|
| - Failed to provide guidance<br>- Failed to provide operational doctrine<br>- Failed to provide oversight<br>- Failed to provide training<br>- Failed to track qualifications<br>- Failed to track performance Planned<br><br>**Inappropriate Operations**<br>- Failed to provide correct data<br>- Failed to provide adequate brief time<br>- Improper manning<br>- Mission not in accordance with rules/regulations<br>- Provided inadequate opportunity for crew rest | - Failed to correct document in error<br>- Failed to identify an at-risk aviator<br>- Failed to initiate corrective action<br>- Failed to report unsafe tendencies<br><br>**Supervisory Violations**<br>- Authorized unnecessary hazard<br>- Failed to enforce rules and regulations<br>- Authorized unqualified crew for flight |

## Planned Inappropriate Operations

Occasionally, the operational tempo and/or the scheduling of aircrew is such that individuals are put at unacceptable risk, crew rest is jeopardized, and ultimately performance is adversely affected. Such operations, though arguably unavoidable during emergencies, are unacceptable during normal operations. Therefore, the second category of unsafe supervision, planned inappropriate operations, was created to account for these failures (Table 8.3).

Take, for example, the issue of improper crew pairing. It is well known that when very senior, dictatorial captains are paired with very junior, weak co-pilots, communication and coordination problems are likely to occur. Commonly referred to as the trans-cockpit authority gradient, such conditions likely contributed to the tragic crash of a commercial airliner into the Potomac River outside of Washington, D.C., in January of 1982. In that accident, the captain of the aircraft repeatedly rebuffed the first officer when the latter indicated that the engine instruments did not appear normal. Undaunted, the captain continued a fatal takeoff in icing conditions with less than adequate takeoff thrust. The aircraft stalled and plummeted into the icy river, killing the crew and many of the passengers.

Clearly, the captain and crew were held accountable. They died in the accident and cannot shed light on causation; but what was the role of the supervisory chain? Perhaps crew pairing was equally responsible. Although not specifically addressed in the report, such issues are clearly worth exploring in many accidents. In fact, in that particular accident, several other training and operating issues were identified.

## Failure to Correct a Known Problem

The third category of known unsafe supervision, Failed to Correct a Known Problem, refers to those instances when deficiencies among individuals, equipment, training or other related safety areas are "known" to the supervisor, yet are allowed to continue unabated (Table 8.3). For example, it is not uncommon for accident investigators to interview the pilot's friends, colleagues, and supervisors after a fatal crash only to find out that they "knew it would happen to him some day." If the supervisor knew that a pilot was incapable of flying safely, and allowed the flight anyway, he clearly did the pilot no favors. The failure to correct the behavior, either through remedial training or, if necessary, removal from flight status, essentially signed the pilot's death warrant – not to mention that of others who may have been on board.

Likewise, the failure to consistently correct or discipline inappropriate behavior certainly fosters an unsafe atmosphere and promotes the violation of rules. Aviation history is rich with by reports of aviators who tell hair-raising stories of their exploits and barnstorming low-level flights (the infamous "been there, done that"). While entertaining to some, they often serve to promulgate a perception of tolerance and "one-up-manship" until one day someone ties the low altitude flight record of ground-level! Indeed, the failure to report these unsafe tendencies and initiate corrective actions is yet another example of the failure to correct known problems.

## Supervisory Violations

Supervisory violations, on the other hand, are reserved for those instances when existing rules and regulations are willfully disregarded by supervisors (Table 8.3). Although arguably rare, supervisors have been known occasionally to violate the rules and doctrine when managing their assets. For instance, there have been occasions when individuals were permitted to operate an aircraft without current qualifications or license. Likewise, it can be argued that failing to enforce existing rules and regulations or flaunting authority are also violations at the supervisory level. While rare and possibly difficult to cull out, such practices are a flagrant violation of the rules and invariably set the stage for the tragic sequence of events that predictably follow.

## Organizational Influences

As noted previously, fallible decisions of upper-level management directly affect supervisory practices, as well as the conditions and actions of operators. Unfortunately, these organizational errors often go unnoticed by safety professionals, due in large part to the lack of a clear framework from which to investigate them. Generally speaking, the most elusive of latent failures revolve around issues related to resource management, organizational climate, and operational processes, as detailed below in Figure 8.5.

**Figure 8.5** Organizational Factors Influencing Accidents

## Resource Management

This category encompasses the realm of corporate-level decision making regarding the allocation and maintenance of organizational assets such as human resources (personnel), monetary assets, and equipment/facilities (Table 8.4). Generally, corporate decisions about how such resources should be managed center around two distinct objectives – the goal of safety and the goal of on time, cost-effective operations. In times of prosperity, both objectives can be easily balanced and satisfied in full. However, as we mentioned earlier, there may also be times of fiscal austerity that demand some give and take between the two. Unfortunately, history tells us that safety is often the loser in such battles and, as some can attest to very well, safety and training are often the first to be cut in organizations having financial difficulties. If cutbacks in such areas are too severe, flight proficiency may suffer, and the best pilots may leave the organization for greener pastures.

Excessive cost cutting could also result in reduced funding for new equipment or may lead to the purchase of equipment that is sub optimal and inadequately designed for the type of operations flown by the company. Other trickle-down effects include poorly maintained equipment and workspaces, and the failure to correct known design flaws in existing equipment. The result is a scenario involving unseasoned, less-skilled pilots flying old and poorly maintained aircraft under the least desirable conditions and schedules. The ramifications for aviation safety are not hard to imagine.

## Culture/Climate

The Organizational Culture/Climate refers to a broad class of organizational variables that influence worker performance. Formally, the terms culture/climate refer to the same thing. An organizational culture can be viewed as the working atmosphere within the organization. One telltale sign of an organization's culture is its structure, as reflected in the chain-of-command, delegation of authority and responsibility, communication channels, and formal accountability for actions (Table 8.4). Just like in the cockpit, communication and coordination are vital within an organization. If management and staff within an organization are not communicating, or if no one knows who is in charge, organizational safety clearly suffers and accidents do happen.

An organization's policies are also good indicators of its culture. Policies are official guidelines that direct management's decisions about such things as hiring and firing, promotion, retention, raises, sick leave, drugs and alcohol, overtime, accident investigations, and the use of safety equipment. *Culture, on the other hand, refers to the unofficial or unspoken rules, values, attitudes, beliefs, and customs of an organization. Culture is "the way things really get done around here."*

**Table 8.4** Selected examples of Organizational Influences (Note: This is not a complete listing)

| Resource / Acquisition Management | Organizational Process |
|---|---|
| ■ Human Resources<br>　■ Selection<br>　■ Staffing / manning<br>　■ Training<br>■ Monetary / budget resources<br>　■ Excessive cost cutting<br>　■ Lack of funding<br>■ Equipment / facility resources<br>　■ Poor design<br>　■ Purchasing of unsuitable equipment<br><br>**Organizational Climate**<br>1. Structure<br>　■ Chain-of-command<br>　■ Delegation of authority<br>　■ Communication<br>　■ Formal accountability for actions<br>2. Policies<br>　■ Hiring and firing<br>　■ Promotion<br>　■ Drugs and alcohol<br>3. Culture<br>　■ Norms and rules<br>　■ Values and beliefs<br>　■ Organizational justice | ■ Operations<br>　■ Operational tempo<br>　■ Time pressure<br>　■ Production quotas<br>　■ Incentives<br>　■ Measurement / appraisal<br>　■ Schedules<br>　■ Deficient planning<br>■ Procedures<br>　■ Standards<br>　■ Clearly defined objectives<br>　■ Documentation<br>　■ Instructions<br>■ Oversight<br>　■ Risk management<br>　■ Safety programs |

When policies are ill defined, adversarial, or conflicting, or when they are supplanted by unofficial rules and values, confusion abounds within the organization. Indeed, there are some corporate managers who are quick to give "lip service" to official safety policies while in a public forum, but then overlook such policies when operating behind the scenes. However, the Third Law of Thermodynamics tells us that, "order and harmony cannot be produced by such chaos and disharmony". Safety is bound to suffer under such conditions.

## Operational Process

This category refers to corporate decisions and rules that govern the everyday activities within an organization, including the establishment and use of standardized operating procedures and formal methods for maintaining checks and balances (oversight) between the workforce and management. For example, such factors as operational tempo, time pressures, incentive systems, and work schedules are all factors that can adversely affect safety (Table 8.4). As stated earlier, there may be instances when those within the upper echelon of an organization determine that it is necessary to increase the operational tempo to a point that overextends a supervisor's staffing capabilities. Therefore, a supervisor may resort to the use of inadequate scheduling procedures that jeopardize crew rest and produce sub optimal crew pairings, putting aircrew at an increased risk of a mishap. However, organizations should have official procedures in place to address such contingencies as well as oversight programs to monitor such risks.

## Conclusion

The Human Factors Analysis and Classification System (HFACS) framework provides a comprehensive, user-friendly tool for identifying and classifying the human causes of aviation accidents. The system, which is based upon Reason's model of latent and active failures, encompasses all aspects of human error, including the conditions of operators and organizational failure. Still, HFACS and any other framework only contributes to an already burgeoning list of human error taxonomies if it does not prove useful in the operational setting.

Given that accident databases can be reliably analyzed using HFACS, the next logical question is whether anything unique will be identified. Early indications within the aviation community suggest that the HFACS framework has been instrumental in the identification and analysis of global human factors safety issues. Consequently, the systematic application of HFACS to the analysis of human factors accident data has afforded the ability to develop objective, data-driven intervention strategies.

Additionally, the HFACS framework and the insights gleaned from database analyses have been used to develop innovative accident investigation methods that have enhanced both the quantity and quality of the human factors information gathered during accident investigations. However, not only are safety professionals better suited to examine human error in the field but, using HFACS, they can now track those areas (the holes in the cheese) responsible for the accidents as well. Only now is it possible to track the success or failure of specific intervention programs designed to reduce specific types of human error and subsequent aviation accidents. In so doing, research investments and safety programs can be either readjusted or reinforced to meet the changing needs of aviation safety.

## Chapter Questions

1. Using the Reason Swiss Cheese Model, identify the correct root cause of the following accidents. The noise and vibration in the cockpit made it difficult to hear and understand speech
   A) Physical Environment
   B) Fitness for Duty
   C) Technological Environment

2. Using the Reason Swiss Cheese Model, identify the correct root cause of the following accidents. Against established policy and management direction, the pilot taxied up to the gate without any ramp personnel to guide him
   A) Decision Error
   B) Routine Violation
   C) Exceptional Violation

3. Using the Reason Swiss Cheese Model, identify the correct root cause of the following accidents. Though the boss knows that a particular pilot is a risk-taker and sometimes pushed the envelope when flying, he did not take action (e.g. by confronting the pilot about his way of flying.)
   A) Failure to correct a known problem
   B) Routine error
   C) Unsafe act

4. Using the Reason Swiss Cheese Model, identify the correct root cause of the following accidents. The pilot misunderstood the instructions from air traffic control and did not seek clarification
   A) Communication and Coordination
   B) Physical limitation
   C) All the above

5. What percentage of aviation accidents can be attributed, in part, to human error?
   A) Approximately 25–50 percent
   B) 70–80 percent
   C) Approximately 50–75 percent

# Accident Investigation Theory

## LEARNING OBJECTIVES

1. Identify the difference between reactive and proactive safety.
2. Know the difference between the NTSB and the FAA.
3. Understand the difference between and aircraft accident and an aircraft incident.
4. Identify the difference between accident definitions from NTSB Part 830 and ICAO Annex 13.
5. Be able to name and understand the three types of aircraft investigations.
6. Name and understand the three types of aircraft accidents.
7. Know and explain the importance of Dr. James Reason's Swiss Cheese accident model.

Reactive safety, accident investigation, is an integral part of the safety process. In the next chapter we will explore the actual process of investigation. In this chapter, we are going to cover the theory behind investigation. This will include the NTSB and ICAO definitions and regulations. We will also look into James Reason's accident causation model and the steps involved with an accident investigation. Why is an accident investigation important? Well, we know that the accident has already occurred. We can't change that fact. But, by investigating the accident we can hopefully prevent the same thing from causing another accident. This type of safety is very complimentary with a proactive safety program, which will be discussed in Chapter 6.

In the United States, the NTSB is the main institution tasked with the investigation of aviation accidents. The NTSB is a very small organization (< 500 employees total). The structure of the organization is seen in Figure 9.1.

### A few facts about the NTSB:

- Entirely independent government agency, reporting directly to the President of the United States.
- Required to determine the probable cause of any accidents related to: civil aviation, highway, railroad, major marine, pipeline.
- Under the Transportation Act of 1974, the board is required to take certain actions during and after an accident investigation.

Because of their relatively small size they do tend to delegate some of the smaller general aviation accidents to the FAA. The FAA will investigate the accident on behalf of the NTSB with the final report being produced by the NTSB, just as if they had completed the investigation themselves. The rules for this fall under NTSB Part 830. According to Part 830, the NTSB is required to take certain steps during an investigation. Part 830.2 defines an accident as:

> "An event associated with the operation of an aircraft that takes place between the time any person boards the aircraft with the intention of flight until such time as all such persons have disembarked, and in which any person suffers death or serious injury, or in which the aircraft receives substantial damage."

In order to really comprehend what this definition means, we need to define what a death (caused by the accident) and serious injury is, and what the definition is of substantial damage to an aircraft.

According to NTSB Part 830.2, in and aircraft accident, death or serious injury is defined as being: " ... *as a result of being in the aircraft, or direct contact with any part of the aircraft, including parts which may have become detached from the aircraft, or exposure to jet blast.*"

### The exceptions being:

> "When injuries are from natural causes, self-inflected, inflicted by other persons, or when the injuries are to stowaways hiding outside the areas normally available to the passengers and crew."

### Notes to remember:

- An injury resulting in death within thirty (30) days of the date of the accident is classified as a fatal injury.
- An aircraft is considered to be missing when the official search has been terminated and the wreckage has not been located.

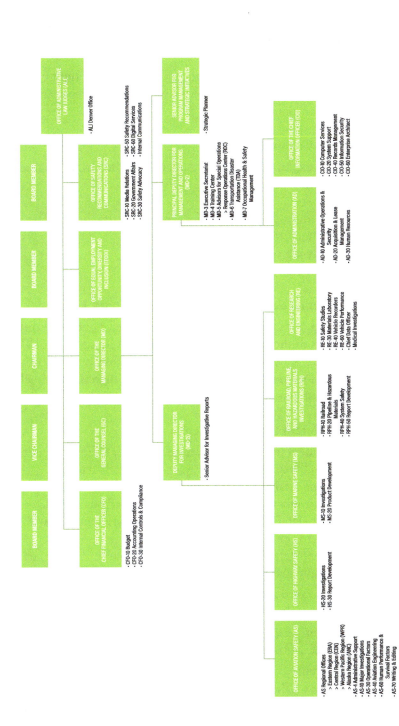

**Figure 9.1** NTSB Organizational Chart

© Kendall Hunt Publishing Company

In order to finish the NTSB definition for an aircraft accident, let's look at what NTBS Part 830 defines substantial damage as:

". . . adversely affects...structural strength, performance, flight characteristics, or...would normally require major repair or replacement of the affected component."

### Exceptions to this definition:

"Engine failure or damage, when the damage is limited to the engine, its cowlings or accessories.

- damage limited to propellers, wing tips, antennas, tires, brakes, fairings, small dents or puncture holes in the aircraft skin.
- the aircraft is missing or inaccessible."

### ICAO has very similar definitions that fall under Annex 13.

These international rules that ICAO has developed have one main difference when it comes to the time limit for a fatality or serious injury. While NTSB Part 830 has a "within 30 days" cutoff for these two definitions, ICAO Annex 13 does not have this cutoff. In fact, there is no timeline listed in Annex 13 at all.

### There are four different types of accidents:

Major accident: accident where aircraft was destroyed, or there were multiple fatalities, or one fatality and the aircraft was substantially damaged.

Serious accident: one fatality without substantial damage to the aircraft, or at least one serious injury and the aircraft is substantially damaged.

Injury accident: non-fatal accident with at least one serious injury without substantial damage to the aircraft.

Damage accident: non-fatal accident without serious injuries, but the aircraft was substantially damaged.

Now that we have looked at the definitions of an aircraft accident, let's look at the definitions of an incident and an occurrence.

Incident: an event other than an accident associated with the operation of an aircraft that affect, or could affect, the safety of operations.

Occurrence: any event not reported as an accident or incident, but still has a significant impact on the overall safety of the operation.

In an accident investigation, the NTSB has a series of protocols that must be followed. The parties to the investigation are one of these protocols. The

NTSB is a very small organization (less than 500 total employees). Because of this, they have to rely on experts from outside the organization to help aid in the investigation. This "party" system takes these experts from the outside organizations that are involved in the accident. The FAA is an automatic party to the investigation, but examples of other parties may be the aircraft manufacturer, the airframe manufacturer, or the powerplant (engine) manufacturer. These organizations, after being named a party to the investigation, would send a representative that has the expertise the NTSB requires to help investigate the accident.

The NTSB has certain rules concerning the representatives of each party. These rules include only gathering the facts pertinent to the accident. There is no here say or opinions allowed during the gathering of facts. The parties are also not allowed to talk to the press or the media. If a party is to break any of the NTSB rules to the investigation, their party status could be revoked or suspended. If the party status is revoked, the organization from which the representative has come from is no longer allowed to be a part of the investigation. If the party status is suspended, that means the organization cannot act as a party to the investigation for a specific period of time.

The FAA will sometimes run its own investigation at the same time the NTSB is conducting its investigation. The purpose of the FAA investigation is to look for blame and usually involves some sort of certificate action, either on the part of the airmen or operator involved. The NTSB is conducting the investigation purely as a way of discovering the safety implications of the accident and trying to prevent the same cause from happening again.

This is because the FAA and the NTSB have vastly different missions. The FAA's mandate is to both promote aviation, along with certificate aviation as well. The NTSB's mission is to find the probable cause and to issue safety recommendations to the FAA and any other pertinent parties to the investigation.

## Types of Investigations

Let's take a look at the three types of NTSB investigations that usually take place after an accident or incident:

- Field Office Investigations
  - investigated by a single field investigator
  - fatal general aviation accidents and some air carrier and commuter accidents with minor injuries

- NTSB Headquarters Investigations
  - Washington, D.C., NTSB Headquarters
  - "Go Team" with an Investigator in Charge or (IIC)

- Specialized Groups for the party system (ATC, weather, avionics, FAA, aircraft manufactures, pilot unions)
- Major air carrier disasters

- Internal Company Investigations
  - Risk Management
  - Evaluate, Educate, Analyze, and Advise

## Types of Accidents

We have looked at accident investigation and the definitions associated with it. Now we will take a look at the three types of accidents.

- Procedural Accidents

Most common, these accidents result from obvious mistakes and generally have a simple, single resolution. Examples include flying into a thunderstorm or taking off with ice or snow on the wings.

- Engineered Accidents

Generally rare accidents with material(s) failures that should have been predicted by designers or discovered by test pilots, but were not. At first, the accident may defy understanding but ultimately yield to examination and result in understandable solutions. Examples include:
  - American Eagle ATR turboprop dives into a frozen field in Indiana, because its de-icing boots did not protect its wings from freezing rain – and as a result new boots are designed, and the entire testing process undergoes review.
  - A TWA Boeing 747-100 explodes off New York because, whatever the source of ignition, its nearly empty center tank contained an explosive mixture of fuel and air.

- System (Organizational) Accidents

These accidents can prove very elusive because of complex organizations. They may involve "contractors" outside the organization and can result from lack of oversight by management or government.
Example includes:

- Valujet DC-9 crash in the Everglades near Miami, FL. The NTSB determined the probable cause to be:

- Failure of SabreTech (contractor) to properly prepare, package, and identify unexpended chemical oxygen generators.
- Failure of ValuJet to properly oversee its contract maintenance, maintenance training and HAZMAT requirements.
- Failure of FAA to require smoke detection & fire suppression in certain cargo (Class D) compartments.

When looking at the different types of accidents, one aspect tends to stand out. Most, if not all, accidents tend to have multiple layers that stacked up to eventually cause the accident. Dr. James Reason developed a model for this. As seen in Figure 9.2, we can see how Dr. Reason's model works.

The "Swiss Cheese Model" shows that accidents can have multiple causes and can be also be more than one type of accident as well. However, this model also shows that if just one layer of the Swiss cheese is removed, the accident may never actually happen.

This chapter has focused on accident theory, including the definitions used by the NTSB and ICAO in identifying accidents and incidents. This is all part of a process called reactive safety. In the next chapter, we will take what we have learned and actually get in to the actual investigation of accidents. This will turn the theory of what has been covered in this chapter, and turn it into the practical.

# BARRIER FAILURE MODE MODEL

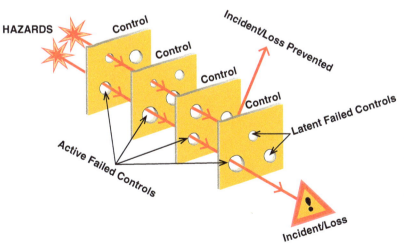

© Kendall Hunt Publishing Company

**Figure 9.2** Reason's Swiss Cheese Model

## Chapter Questions

1. Reactive safety:
   A) Solves the problem before an accident occurs
   B) Investigates the accident after it happens
   C) Is not really necessary
   D) None of the above
2. The three types of accidents include
   A) Engineered accidents
   B) System Accidents
   C) Procedural accidents
   D) All of the above
3. Dr. James Reason's Swiss Cheese accident model shows
   A) Accidents usually have one cause
   B) Are the pilot's fault
   C) Have multiple layers of causes
   D) None of the above
4. Reactive and proactive safety are complimentary to each other:
   A) True
   B) False

# Aircraft Accident Investigation

## Site Safety at the scene of the accident

An aviation accident often causes injuries and loss of life. We do not want the investigation of this horrible event to cause injuries or death to the accident investigators. The application of safety management in the conduct of aircraft accident investigation operations is complex. There are a range of factors that have a significant effect on the safety management process. Unlike personnel involved in the more predictable domains within the aviation industry, investigators are required to respond to accident situations that are variable in nature, scale, and environment. These factors make the identification of hazards and determination of exposure a more difficult exercise. Furthermore, given the relative infrequency of accidents, there are few opportunities for the scientific analysis of aircraft debris that is essential for accurate assessment of occupational health risks.

The need for prior planning and training cannot be overemphasized, especially during the initial investigation when critical accident information can be easily lost or contaminated. The collection of fluid samples from various systems is time critical, while haphazard collection of samples will cause contamination and provide misleading indication of system deficiencies. Flight control actuators, flap and control surface deflections and cockpit switch locations are all critical evidence that must be properly documented as rapidly as possible without causing additional hazards to the investigators.

A wide range of control measures can be applied to help reduce risks, including:

- stopping or delaying the task – where risk is shown to be excessive, this may be the only option until alternative methods of work are established;
- removal/isolation of hazards – components can be disconnected, made safe or removed from the site, hazardous materials can be neutralized or covered, dust and fibers can be suppressed with water or fluids, etc.;
- limiting exposure – reduce the numbers of personnel within hazardous areas or limit the length of time or frequency of exposure;
- modifying tasks or using alternative equipment or materials – this course of action can produce significant reductions in risk;
- employing specific work procedures (e.g. exposure control plans); and f) using protective clothing/equipment.

A hazard is something that has the potential to cause adverse consequences, and the degree of adverse consequences caused by specific exposures is important when determining the risk posed. A wide range of hazards may exist at aircraft accident sites, some of which may not be directly associated with the aircraft wreckage. Hazards may be posed by pathogens (from human or animal remains), cargo, and the nature of the accident location, ground installations, and other factors. Given the wide range of potential hazards at an accident site, it can be helpful to categorize typical hazards, in order to better manage the accident site. Hazards have been categorized as follows:

- Environment – location (both geographic and topographic), fatigue (effects of travel and transportation), insects/wildlife, climate, security and political situation;
- Physical – fire, stored energy, explosives, structures;
- Biological – pathogens associated with human remains or cargo consignments and state of local hygiene;
- Materials – exposure to and contact with materials and substances at the site; and
- Psychological – stress and traumatic pressures imposed by exposure to the aircraft accident, and interaction with those associated with the air carrier and related aviation activities.

## Site Survey

The investigation at the accident site should begin with an assessment of the wreckage. Accident investigators begin with the 4-corners search. The four corners of the aircraft are the left wingtip to the right wingtip; the nose of the aircraft all the way back to the tail of the aircraft. During this initial, cursory, examination of the site determine that the major structural members are present; wings, vertical and horizontal tail, the correct number of engines, the correct number of propellers and propeller blades, etc. Every piece of debris is important, however pay particular attention ensuring that major structural and flight control surfaces are within the wreckage pattern. This initial survey must determine whether all flight control surfaces are present as well; ailerons, flaps, elevators, trim tabs, spoilers, etc. As the number of surfaces may be quite extensive, a common practice is to have each member possess a simple diagram of the aircraft (usually obtainable from the Operator's or Maintenance Manual). As each structural section is identified and each flight control surface is found, the appropriate part of the illustration can be "colored in". Later, all illustrations can be compared to assure the investigation team that the entire aircraft it at the site. The lack of a major section, or control surface may be indicative of a loss prior to impact and the effort to recover the missing parts should begin as soon as possible. Hence, the need to accomplish this basic inventory early in the investigation.

An assessment of the basic terrain features surrounding the accident site should be made prior to detailed analysis. If the terrain is rising as when the impact occurred, the evidence of impact may indicate a steeper angle than would exist if the terrain were level or descending. Similarly, if the area is heavily forested, the degree of impact may be greater than if the area were devoid of large structures or vegetation.

The investigators should determine the scope of the aircraft breakup. If extensive, this may be accomplished by a site walk-through. Beginning at the point of initial contact with the ground the investigator should attempt to determine a basic breakup direction and begin walking that line. Identification of structure or parts along the path can be made with noting whether they are straight ahead or to the left or right. A preliminary sketch of the wreckage found might be made without great effort as to scale. When the last parts are noted in the line walked, it might be assumed that no other parts exist further down. The investigator should continue the line for some distance assuring that what was suspected to be the last parts are, in fact, so. Frequently, heavy objects having great inertia may lie well beyond the normal wreckage pattern. Once the length boundary is determined, the investigator should similarly determine the width to either side. The importance is that all investigative efforts can then be concentrated within an established boundary. This facilitates investigative assignments and assures that investigation team members do not stray beyond the boundaries without coordination with the investigator-in-charge. In the case of in-flight breakup or mid-air collision, radar data may be useful in identification of the initial debris field and extent of wreckage disbursement. This will also be vital in over water accident site determination.

## Wreckage patterns

After the initial study of the general scene of the accident has been made and photographs taken the first step in the actual investigation is usually that of plotting the distribution of the wreckage. In simple terms this is done by measuring the distances and bearings of the main wreckage and also of the scattered parts of the wreckage, including the contents of the aircraft, survivors and victims, all impact and ground markings, and then recording this information on a chart to a convenient scale. In many accidents the preparation of a wreckage distribution chart is a task considered to be well within the capabilities of an investigator. If a GPS plotting has not been accomplished, consideration should be given to employing the services of a qualified surveyor when the circumstances of the accident are such that there has been extensive scattering of the wreckage. The preparation of a wreckage distribution chart is worthy of painstaking effort to ensure its completeness and accuracy, for the study of the completed chart may suggest possible failure patterns or sequences, and the significance of later findings may often depend upon reference to the original chart. It will not only be used as a reference document throughout the investigation but also it will remain a most important document for inclusion in the investigator's dossier and will supplement the written report. In determining the type and amount of information to be included on the chart of any specific accident, the investigator must be guided by the circumstances surrounding the particular accident. In most cases the chart should record the locations of all major components, parts and accessories, freight, and the locations at which the accident victims were found, or survivors located, and if available, their identities. The initial contact markings and other ground markings should also be indicated on the chart with suitable reference to identify the part of the aircraft or component responsible for the marking. When terrain features appear to have a bearing on the accident or on the type or extent of structural damage, they too should be noted on the wreckage distribution chart. Pertinent dimensions, descriptive notes and the locations from which photographs were taken add to the completeness of the chart. The preparation of a wreckage distribution sketch may be accomplished in various ways but the following are some examples of simple methods:

- When the wreckage is concentrated in a small area, distances and bearings (magnetic) can be measured from a central point of the wreckage. The plotting of the items can be made on a polar diagram.

- When the wreckage is scattered, a base line can be laid out usually along the main wreckage trail, dependent upon the terrain, and distances measured along the base line from a reference point and then perpendicularly from the base line to the scattered pieces of wreckage. A chart is then prepared from this information using a suitable scale. The use of squared paper may be useful in preparing simple plots.

Where there are many pieces of wreckage the presentation of the chart can be simplified by using a letter or a numeral for each item and preparing a suitable index for inclusion on the chart.

## Examination of ground scars

The marks of first impact of the aircraft with the ground should be found. From these and the distribution of the wreckage, it can usually be determined which part of the aircraft struck the ground first. The path of the aircraft may be deduced by careful examination of ground marks or scars upon trees, shrubs, rocks, poles, power lines, buildings, etc. Wing tips, propellers or landing gear leave telltale marks or torn-off parts at points of contact with fixed objects. Ground scars used in conjunction with height of broken trees or brush will assist in establishing the angle and attitude in which the aircraft struck the ground. Examination of the victims of the accident and the contents of the aircraft, can also assist in establishing angle, attitude, and speed at impact. The general state of distortion and "telescoping" of the structure will permit an investigator to deduce whether the aircraft crashed at high or low speed. Usually only local damage occurs at low speed impact, but at high-speed wings and tail become buckled and foreshortened. Cases have occurred in which the aircraft has been completely buried in a deep crater, with only a few twisted fragments dispersed adjacent to the impact site. Short straight furrows running out from each side of the crater told where the leading edges of the outer wings had hit the ground while traveling almost vertically downwards at very high speed. When engines have not penetrated into the ground their vertical descent speed has probably been small, but the aircraft might have been traveling very fast at a shallow angle and in such circumstances, the wreckage will be spread far along a line from the mark of first impact. If the wreckage is widely scattered along the flight path, this may indicate that some structural disintegration had occurred before impact with the ground. It is usually possible to form a preliminary mental picture of:

- the direction, angle and speed of descent;
- whether it was a controlled or uncontrolled descent;
- whether the engines were under power at the time of impact;
- whether the aircraft was structurally intact at the point of first impact.

The extent of the damage to the wreckage will give some preliminary indication of the evidence that can be obtained from it by subsequent detailed examination. If structural disintegration in the air is suspected, it is essential to plan the investigation to ensure that all information, which will help to trace the primary failure, is extracted from the wreckage before it is moved. In such circumstances aircraft wreckage may be scattered over several miles of woodland, field, marsh, or built-up area and may be difficult to locate. Search parties should comb the district and the search should be continued until all significant components have been found. The cooperation in the search of military personnel, police, schools and local residents should be requested but at the same time searchers should be informed of the need to report the location of pieces of wreckage without disturbing them. This will enable the investigator to examine and determine the exact location of such pieces as they fell to the ground. Light detached portions of low density tend to drift in the direction of the prevailing wind at the time of the accident whilst wind effects will less affect dense objects, and knowledge of this direction may save time in locating aircraft pieces. No piece of wreckage should be disturbed or removed until:

- its position is recorded;
- an identification number is painted on it on an undamaged area, or in the case of small portions, a label attached; and
- notes are made of the manner in which the piece struck the ground, what the nature of the ground was, and whether it hit trees or buildings, etc., prior to this.

Such notes and photographs will be very valuable when a later detailed examination is made and may help to separate ground impact damage from the other damage. A special search should be made for any part of the aircraft not accounted for at the accident site and if it cannot be located, the fact must be recorded. In the case of accidents associated with wheels-down landings, tire marks should be carefully recorded and examined. The width of the tire imprint of each wheel and the density of the color of the marks should be noted. The tire marks may well provide evidence of braking or skidding or sliding and, in particular, may provide a clue to a hydro-planing situation. A hydro-planing tire may leave a very distinctive whitish mark on the runway. These tracks are the result of a scrubbing action which is provided by the forces under the tire during hydro-planing. It should never be overlooked that the victims of an aircraft accident, if objectively examined in the same manner as the aircraft wreckage, may reveal important information relating to aircraft speed, aircraft attitude at impact, sequence of break-up, etc. This is referred to in more detail in the Human Factors Investigation.

## Impact marks

Impact marks, both on the ground and on the wreckage, can provide valuable information to understand the aircraft attitude and path at impact. Impact marks (also referred to as witness marks and ground scars) should be thoroughly measured and documented as soon as possible, as they are often obscured during the investigation. Skid marks and tire tracks, of course, are particularly useful for overruns and runway excursions.

Witness marks on analog instrument readings at the time of impact have been a traditional source of useful information. Many digital instruments have memory circuits, or illuminated portions that may record impact conditions and their circuitry needs to be handled carefully until it can be analyzed in the laboratory.

## Need for photographs

Photography is an important element of the investigation process. Clear, well composed photographs allow the investigator to preserve perishable evidence, substantiate the information in the report, and illustrate the investigator's conclusions. Every accident investigator needs a basic knowledge of photography. This allows the investigator to take quality photographs or to communicate effectively with a professional photographer in order to obtain photographs that contribute to a clearly written report.

## What to photograph at the accident site

The general rule in accident site photography is to start with the most perishable evidence and work to the least perishable evidence. The following is an example of an investigator's checklist for accident site photography.

## Immediately following the mishap.

- Firefighting (Video)
- Rescue activities (Video)
- Radar and ATC (Recordings)
- Weather (forecast and actual conditions)

## Once the investigation begins.

- Aerial view of the site (Video)
- The site ground view from each cardinal compass position
- The site from the direction the aircraft was traveling at impact
- Ground scars
- Damage to trees and foliage

- Skid marks
- Photo inventory of major wreckage components
- Flight control surfaces and actuators
- Landing gear and other hydraulic components
- Cockpit switch positions
- Fire/heat damage and discoloration
- Human remains, injuries, blood/tissue smears on wreckage
- Extra items or items adjacent to items not accounted for
- Close-ups of fracture surfaces
- Close-ups of improperly installed components
- Close-ups of any other items you suspect may have contributed to the mishap
- Private property damage
- Steps in removing, opening or cutting apart components
- Any other photos deemed necessary

Begin by using a video camera to record firefighting and rescue activities. Place the camcorder on a tripod and zoom the lens back to cover the whole site. Turn it on and let it record continuously. For a response that takes a long period of time, be ready to change tapes as they are consumed. If you have more than one video camera, get video from as many vantage points as possible without interfering with the response activities. The video will be valuable later for a number of uses. It will provide a record of the response. Investigators can use the video to determine what damage responders caused and what damage was caused by the accident itself. The video can be used to train firefighting and rescue crews. If you can do so without endangering yourself or interfering with the response, photograph other perishable items of evidence such as ground scars and skid marks.

As soon as the fire is extinguished and the accident site is declared safe for investigators to enter the area, photograph any remaining skid marks and ground scars. The local medical examiner will begin removing the human remains, if any, from the scene. All pieces of human remains should be photographed and cataloged before they are moved. Any other medical evidence such as tissue smears on wreckage should be photographed as soon as possible. Additionally, document any damage to private property.

The next step is to take aerial photographs of the site. An easy and effective way to accomplish this is to use a contractor who specializes in aerial photography. Many times, the use of an unmanned aerial system can give you high definition video and pictures quickly. Additionally, the use of unmanned aerial systems often allow coaching for the pilot to ensure the proper information is captured. If you hire an aerial photographer, make sure the aircrew understands exactly what information you need to capture.

If you take the photographs yourself, an effective way is to use a helicopter. When taking photographs from a helicopter, remember to secure yourself and your equipment. Hold the camera so that the lens axis is as near to the vertical as possible. Take several photos from each position using different exposures. Use a high shutter speed to obtain the clearest images. Do not allow the camera to touch the helicopter's structure during exposure, because this will allow the vibrations from the helicopter to be transmitted to the camera, degrading the clarity of the image. Have a piece of equipment, such as a vehicle, in the photos to provide a sense of scale. If possible, take aerial photos at different times of the day. The different shadow patterns will reveal different details.

During the initial walkthrough of the accident site, it may be helpful to have an assistant carry a video camcorder. Record initial impressions as video notes with a voice accompaniment. This can also be used as a briefing tool for newly arriving members of the investigation team, and to hazardous material mitigation crews and medical personnel for the recovery of fatal passengers. The next major photography task is to photograph the wreckage. If the wreckage is concentrated in a small area and all of it is easily seen from a single vantage point, photograph it from all cardinal and intermediate compass points. The photographer should stand the same distance from the center of the wreckage while taking each photograph. Be sure your notes reflect the direction you were facing at the time. If the wreckage is spread out over a large area, it may not be practical to photograph the whole scene. In this case, photograph each significant piece or group of pieces of wreckage.

As a minimum, take a photograph from all cardinal compass points, then move closer to show details. Be sure to note the location of the part of group of wreckage on the accident diagram. Take photos that illustrate damage to the components, fracture surfaces, and witness marks. The photographer or investigator should never try to reassemble broken parts as this may destroy the fracture surface and disturb the evidence of the cause of the failure. When the wreckage is removed from the site or if it is moved to provide access to other evidence, be sure to photograph it before it is disturbed. Any time major pieces of wreckage are moved from the site, use a video recorder to record the process of preparing them for transport, loading them onto the vehicle, and removing and setting them up at the destination.

Whenever components are dismantled or cut open, record the process on video if possible. Other significant pieces of evidence to be photographed include evidence of fire, heat discoloration of structures, structural fractures, switch positions, and circuit breakers. Any damage to nearby trees and foliage should be photographed, as well as ground scars from pieces of wreckage after the aircraft initial breakup. Photograph the impact point from a vantage point that is along the flight path of the aircraft. Anything found in the wreckage that should not be there should be photographed.

Also, photograph anything that has a critical component missing. For instance, if the investigation reveals a missing cotter pin on a critical fastener, photograph that fastener and photograph one that shows a normal installation.

Environmental conditions should also be documented if there is any consideration that weather, sun angle, visual illusion or lack of visual reference as a possible contribution of the accident. This would require not only the documentation of weather conditions as soon as possible at the time of the accident, but to recreate the sun or moon angle and conditions at another date under the same representative conditions. This may also be accomplished by simulation, especially when controlled flight into terrain is being investigated, so as not to hazard another aircraft in attempting to recreate the accident conditions.

Considering the cost of an aircraft accident and its investigation, photography is inexpensive. Take as many pictures as needed. Take one photo at the normal exposure, then take the same subject at half than double the cameras indicated exposure. Photographers call this technique "bracketing." It ensures at least one photograph will be properly exposed. Move your flash attachment if, you are using one, to have the subject illuminated from several different directions. Take notes along with the photographs. The notes must contain enough information to later identify each photograph and its significance. This is not as important when using digital camera as each photograph can be reviewed immediately to see if they contain the necessary information.

## Aircraft Accident Working Groups

The accident investigation team's immediate boss is the Investigator-in-Charge (IIC), a senior investigator with years of NTSB and industry experience. Each investigator is a specialist responsible for a clearly defined portion of the accident investigation. In aviation, these specialties and their responsibilities are:

OPERATIONS: The history of the accident flight and crewmembers' duties for as many days prior to the crash as appears relevant.

STRUCTURES: Documentation of the airframe wreckage and the accident scene, including calculation of impact angles to help determine the plane's pre-impact course and attitude.

POWERPLANTS: Examination of engines (and propellers) and engine accessories.

SYSTEMS: Study of components of the plane's hydraulic, electrical, pneumatic and associated systems, together with instruments and elements of the flight control system.

AIR TRAFFIC CONTROL: Reconstruction of the air traffic services given the plane, including acquisition of ATC radar data and transcripts of controller-pilot radio transmissions.

WEATHER: Gathering of all pertinent weather data from the National Weather Service, and sometimes from local TV stations, for a broad area around the accident scene.

HUMAN PERFORMANCE: Study of crew performance and all before-the-accident factors that might be involved in human error, including fatigue, medication, alcohol. Drugs, medical histories, training, workload, equipment design and work environment.

SURVIVAL FACTORS: Documentation of impact forces and injuries, evacuation, community emergency planning and all crash-fire-rescue efforts.

Under direction of the IIC, each of these NTSB investigators heads what is called a "working group" in one area of expertise. Each is, in effect, a subcommittee of the overall investigating team. The groups are staffed by representatives of the "parties" to the investigation (see the next section – The Party System) – the Federal Aviation Administration, the airline, the pilots' and flight attendants' unions, airframe and engine manufacturers, and the like. Pilots would assist the operations group; manufacturers' experts, the structures, systems and power plants groups; etc. Often, added groups are formed at the accident scene – aircraft performance, maintenance records, and eyewitnesses, for example. Flight data recorder and cockpit voice recorder teams assemble at NTSB headquarters.

In surface accident investigations, teams are smaller and working groups fewer, but the team technique is the same. Locomotive engineers, signal system specialists and track engineers head working groups at railroad accidents. The specialists at a highway crash include a truck or bus mechanical expert and a highway engineer. The Board's weather, human performance and survival factors specialists respond to accidents of all kinds.

At least once daily during the on-scene phase of an investigation, one of the five Members of the Safety Board itself, who accompanies the team, briefs the media on the latest factual information developed by the team.

While a career investigator runs the inquiry as Investigator-in-Charge, the Board Member is the primary spokesperson for the investigation. A public affairs officer also maintains contact with the media. Confirmed, factual information is released. There is no speculation over cause.

The individual working groups remain as long as necessary at the accident scene. This varies from a few days to several weeks. Some then move on – power plants to an engine teardown at a manufacturer or overhaul facility; systems to an instrument manufacturer's plant; operations to the airline's training base, for example. Their work continues at Washington headquarters, forming the basis for later analysis and drafting of a proposed report that goes to the Safety Board itself perhaps 12 to 18 months from the date of the accident. Safety recommendations may be issued at any time during the course of an investigation.

## Parties to the investigation

The Board investigates about 2,000 aviation accidents and incidents a year, and about 500 accidents in the other modes of transportation – rail, highway, marine and pipeline. With about 400 employees, the Board accomplishes this task by leveraging its resources. One way the Board does this is by designating other organizations or companies as parties to its investigations.

The NTSB designates other organizations or corporations as parties to the investigation. Other than the FAA, which by law is automatically designated a party, the NTSB has complete discretion over which organizations it designates as parties to the investigation. Only those organizations or corporations that can provide expertise to the investigation are granted party status and only those persons who can provide the Board with needed technical or specialized expertise are permitted to serve on the investigation; persons in legal or litigation positions are not allowed to be assigned to the investigation. All party members report to the NTSB.

Eventually, each investigative group chairman prepares a factual report and each of the parties in the group is asked to verify the accuracy of the report. The factual reports are placed in the public docket.

## Chapter Questions

1. List the 5 important hazards discussed in this chapter.
2. Explain the 4-corners concept.
3. Identify two reasons to use an unmanned aerial vehicle to photograph an accident site.
4. What are the names of the working groups used to investigate major aircraft accidents?

# Aviation Safety Program Management and Safety Management System (SMS)

**11**

## LEARNING OBJECTIVES

1. Understand the 4 components of Safety Management Systems (SMS).
2. Explain the differences between a good safety program and SMS.
3. Understand the importance of creating a robust safety policy.
4. List the important characteristics of Safety Risk Management.
5. List the important characteristics of Safety Assurance.
6. List the important characteristics of Safety Promotion.

## Safety Management Systems (SMS)

The implementation of a safety management system (SMS) represents a fundamental shift in the way the organization does business. Safety management systems require organizations to adopt and actively manage the elements detailed in this document and to incorporate them into their everyday business or organization practices. In effect, safety becomes an integral part of the everyday operations of the organization and is no longer considered an adjunct function belonging to the safety office.

The word system means "to bring together or combine." This is not a new term. The philosopher Aristotle first identified systems. SMS involves the transfer of some of the responsibilities for aviation safety issues from the regulator to the individual organization. This is a role shift in which the regulator oversees the effectiveness of the safety management system but withdraws from day-to-day involvement in the organizations it regulates. The day-to-day issues are discovered, analyzed and corrected internally by the organizations.

From the organization's perspective, the success of the system hinges on the development of a safety culture that promotes open reporting through non-punitive disciplinary policies and continual improvement through proactive safety assessments and quality assurance.

The safety management system philosophy requires that responsibility and accountability for safety be retained within the management structure of the organization. Management is ultimately responsible for safety, as they are for other aspects of the enterprise. The responsibility for safety, however, resides with every member of the organization. In safety management, everyone has a role to play. Regardless of the size and complexity of an organization, senior management will have a significant role in developing and sustaining an organization safety culture. Without the sincere, unconditional commitment of all levels of management, any attempt at an effective safety program will be unsuccessful. Safety management requires the time, financial resources and consideration that only the senior management can provide.

Some examples of management commitment and support may be: discussing safety matters as the first priority during staff meetings, participating in safety committees and reviews, allocating the necessary resources such as time and money to safety matters, and setting a personal example. However it is manifested, the importance of support from management cannot be underestimated.

## Why is SMS Needed?

SMS facilitates the proactive identification of hazards, promotes the development of an improved safety culture, modifies the attitudes and behavior of personnel in order to prevent damage to aircraft or equipment, as well as makes for a safer workplace. SMS helps organizations avoid wasting financial and human resources, in addition to wasting management's time from being focused on minor or irrelevant issues.

SMS allows employees to create ownership of the organizational process and procedures to prevent errors. SMS lets managers identify hazards, assess risk and build a case to justify controls that will reduce risk to acceptable levels. SMS is a proven process for managing risk that ties all elements of the organization together, laterally and vertically, and ensures appropriate allocation of resources to safety issues. An SMS provides an organization with the capacity to anticipate and address safety issues before they lead to an incident or accident. An SMS also provides management with the ability to deal effectively with accidents and near misses so that valuable lessons are applied to improve safety and efficiency. The SMS approach reduces loss and improves productivity.

## Definition of an SMS

SMS is defined as a coordinated, comprehensive set of processes designed to direct and control resources to optimally manage safety. SMS takes unrelated processes and builds them into one coherent structure to achieve a higher level of safety performance, making safety management an integral part of overall risk management. SMS is based on leadership and accountability. It requires proactive hazard identification, risk management, information control, auditing and training. It also includes incident and accident investigation and analysis. Figure 11.1 contrasts the attributes of a successful SMS vs. the attributes of a good safety program.

Safety management is woven into the fabric of an organization. It becomes part of the culture – the way people do their jobs. The organizational structures and activities that make up a safety management system are found throughout an organization. Every employee contributes to the safety health of the organization. In some organizations, safety management activity will be more visible than in others, but the system must be integrated into "the way things are done." This will be achieved by the implementation and continuing support of a safety program based on coherent policies and procedures.

## The Accountable Executive

One person must have the responsibility to oversee SMS development, implementation and operation. This person is called the Accountable Executive. The accountable executive must be the "champion" for the SMS program. The managers of the "line" operational functions, from middle management

**Figure 11.1** Attributes of an SMS

to front-line managers and supervisors, manage the operations in which risk is incurred. These managers and supervisors are the "key safety personnel" of the SMS. For each process, the element that defines responsibilities for definition, and documentation of aviation safety responsibilities, applies to all components, elements and processes.

## Key Safety Personnel

Top management has the ultimate responsibility for the SMS and should provide the resources essential to implement and maintain the SMS. Top management should appoint members of management, who, irrespective of other responsibilities, have responsibilities and authority including:

- Ensuring the processes needed for the SMS are established, implemented and maintained.
- Ensuring the promotion of awareness of safety requirements throughout the organization.
- Ensuring that aviation safety-related positions, responsibilities, and authorities are defined, documented and communicated throughout the organization.

## The Four Components, or Pillars, of SMS:

The ICAO Document 9859 and FAA Advisory Circular 120-92B states that SMS is structured upon four basic components, or sometimes called pillars, of safety management:

- Safety Policy
- Safety Risk Management
- Safety Assurance
- Safety Promotion

## Safety Policy

Every type of management system must define policies, procedures and organizational structures to accomplish its goals. An SMS must have policies and procedures in place that explicitly describe responsibility, authority, accountability and expectations. Most importantly, safety must be a core value.

The safety policy should state that safety has a very high priority within the organization. It is the accountable manager's way of establishing the importance of safety as it relates to the overall scope of operations. Leadership sets the tone. Senior management commitment will not lead to positive action unless commitment is expressed as direction. Management must develop and communicate safety policies that delegate specific responsibilities and hold people accountable for meeting safety performance goals.

The policy must be clear, concise and emphasize top level support – including a commitment to:

- Implementing an SMS
- Continuous improvement in the level of safety
- Managing safety risks
- Complying with applicable regulatory requirements
- Encouragement of, not reprisal against, employees that report safety issues
- Establishing standards for acceptable behavior
- Providing management guidance for setting and reviewing safety objectives
- Documentation
- Communication with all employees and parties
- Periodic review of policies to ensure they remain relevant and appropriate to the organization
- Identifying responsibility of management and employees with respect to safety performance
- Integrating safety management with other critical management systems within the organization
- Safety component to all job descriptions that clearly defines the responsibility and accountability for each individual within the organization

## Safety Risk Management

A formal system of hazard identification and management is fundamental in controlling an acceptable level of risk. A well-designed risk management system describes operational processes across department and organizational boundaries, identifies key hazards and measures them, methodically assesses risk, and implements controls to mitigate risk.

Understanding the hazards and inherent risks associated with everyday activities allows the organization to minimize unsafe acts and respond proactively, by improving the processes, conditions and other systemic issues that lead to unsafe acts. These systemic/organizational elements include – training, budgeting, procedures, planning, marketing and other organizational factors known to play a role in many systems-based accidents. In this way, safety management becomes a core function and is not just an adjunct management task. It is a vital step in the transition from a reactive culture, one in which the organization reacts to an event, or to a proactive culture, in which the organization actively seeks to address systemic safety issues before they result in an active failure. The fundamental purpose of a risk management system is the early identification of potential problems.

The risk management system enhances the manner in which management safety decisions are made. The risk management process identifies the 6 steps outlined below:

- Establish the Context. This is the most significant step of the risk process. It defines the scope and definition of the task or activity to be undertaken, the acceptable level of risk is defined, and the level of risk management planning needed is determined.
- Identify the Risk. Identification of what could go wrong and how it can happen is examined, hazards are also identified and reviewed, and the source of risk or the potential causal factors are also identified.
- Analyze the Risk. Determine the likelihood and consequence of risk in order to calculate and quantify the level of risk. A good tool for this process is the reporting system for information gathering technique. Determining the frequency and consequence of past occurrences can help to establish a baseline for your risk matrix. Each organization will have to determine their definition of severity according to its individual risk aversion.
- Evaluate the Risks. Determine whether the risk is acceptable or whether the risk requires prioritization and treatment. Risks are ranked as part of the risk analysis and evaluation step.
- Treat the Risks. Adopt appropriate risk strategies in order to reduce the likelihood or consequence of the identified risk. These could range from establishing new policies and procedures, reworking a task, or a change in training, to giving up a particular mission or job profile.
- Monitor and Review. This is a required step at all stages of the risk process. Constant monitoring is necessary to determine if the context has changed and the treatments remain effective. In the event the context changes, a reassessment is required.

## Risk Assessment

Risk assessment is a decision step, based on combined severity and likelihood. Ask; is the risk acceptable? The risk assessment may be concluded when potential severity is low or if the likelihood is low or well controlled.

## Risk Matrix

The risk assessment matrix is a useful tool to identify the level of risk and the levels of management approval required for any Risk Management Plan. There are various forms of this matrix, but they all have a common objective to define the potential consequences and/or severity of the hazard versus the probability or likelihood of the hazard.

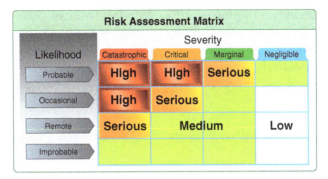

**Figure 11.3** Risk Matrix

*Source:* FAA.gov

To use the risk assessment matrix effectively it is important that everyone have the same understanding of the terminology used for probability and severity. For this reason, definitions for each level of these components should be provided. Figure 11.3 shows a risk matrix used by many aviation organizations.

## Risk Control

Often, risk mitigation will require new processes, new equipment or changes to existing ones. Look at the system with the proposed control in place to see if the level of risk is now acceptable.

Stay in this design loop until it is determined that the proposed operation, change, etc. not be mitigated to allow operations within acceptable levels of risk.

## Safety Assurance

Policies, process measures, assessments and controls are in place. The organization must incorporate regular data collection, analysis, assessment and management review to assure safety goals are being achieved. Solid change management processes must be in place to assure the system is able to adapt.

The ongoing monitoring of all systems and the application of corrective actions are functions of the quality assurance system. Continuous improvement can only occur when the organization displays constant vigilance regarding the effectiveness of its technical operations and its corrective actions. Without ongoing monitoring of corrective actions, there is no way of telling whether the problem has been corrected and the safety objective met. Similarly, there is no way of measuring if a system is fulfilling its purpose with maximum efficiency. Evaluation of the safety program includes external assessments by professional or peer organizations. Safety oversight is provided in part by some of the elements of the SMS such as occurrence reporting and investigation. However, safety assurance and oversight programs proactively seek out potential hazards based on available data as well as the evaluation of the organization's safety program. This can best be accomplished by:

- Conducting internal assessments of operational processes at regularly scheduled intervals.
- Utilizing checklists tailored to the organization's operations when conducting safety evaluations.
- Assessing the activities of contractors where their services may affect the safety of the operation.
- Having assessment of evaluator's processes conducted by an independent source.
- Documenting results and corrective actions.
- Documenting positive observations.
- Categorizing findings to assist in prioritizing corrective actions.
- Sharing the results and corrective actions with all personnel.
- Utilizing available technology such as Health Usage Monitoring Systems (HUMS) to supplement quality and maintenance programs and Flight Data Monitoring to evaluate aircrew operations.
- Facilitating Safety Committee meetings.
- Advising the CEO (Accountable Executive) on safety issues.
- Causing incidents to be investigated and reviewed, making recommendations and providing feedback to the organization.
- Conducting periodic assessment of flight operations.
- Providing safety insight to the organization's management.

Monitoring by audit forms a key element of this activity and should include both a quantitative and qualitative assessment. The results of all safety performance monitoring should be documented and used as feedback to improve the system.

It is widely acknowledged that accident rates are not an effective measurement of safety. They are purely reactive and are only effective when the accident rates are high enough. Furthermore, relying on accident rates as a safety performance measurement can create a false impression; an assumption that zero accidents indicate the organization is safe. A more effective way to measure safety might be to address the individual areas of concern. For example, an assessment of the improvements made to work procedures might be far more effective than measuring accident rates.

## Interfaces in Safety Risk Management (SRM) and Safety Assurance (SA)

Safety Risk Management (SRM) and Safety Assurance (SA) are the key functional processes of the SMS. They are also highly interactive. The flowchart on figure 11.4 may be useful to help visualize these interactions. The interface element concerns the input-output relationships between the activities in the processes. This is especially important where interfaces between processes involve interactions between different departments, contractors, etc. Assessments of these relationships should place special attention to flow of authority, responsibility and communication, as well as procedures and documentation.

## Safety Promotion and Safety Culture

The organization must continually promote, train and communicate safety as a core value with practices that support a sound safety culture.

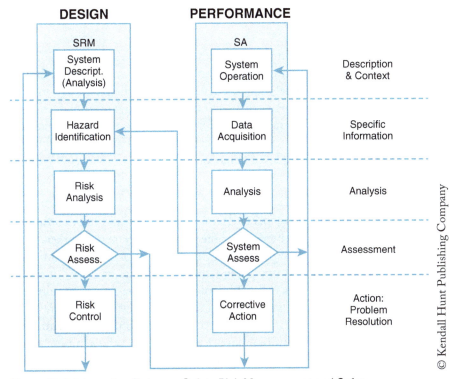

**Figure 11.4** Interaction Between Safety Risk Management and Safety Assurance

An organization's Safety Culture influences the values, beliefs and behaviors that we share with other members of our various social groups. Culture serves to bind us together as members of groups and to provide clues as to how we behave in both normal and unusual situations. Some people see culture as the "collective programming of the mind." Culture is the complex, social dynamic that sets the rules of the game, or the framework for all our interpersonal interactions. It is the sum total of the way people work. Culture provides a context in which things happen. For safety management, understanding the culture is an important determinant of human performance and its limitations. The ultimate responsibility for safety rests with the management of the organization. Safety Culture is affected by such factors as:

- Management's actions and priorities
- Policy and procedure
- Supervisory practices
- Safety planning and goals
- Actions in response to unsafe behaviors
- Employee training and motivation
- Employee involvement or buy-in

An organizational culture recognizes and identifies the behavior and values of particular organizations. Generally, personnel in the aviation industry enjoy a sense of belonging. They are influenced in their day-to-day behavior by the values of their organization. Does the organization recognize merit, promote individual initiative, encourage risk taking, tolerate breaches of SOP's, promote two-way communications, etc.? The organization is a major determinant of employee behavior.

## Positive Safety Culture

A positive safety culture is generated from the "top down." It relies on a high degree of trust and respect between workers and management. Workers must believe that they will be supported in any decisions made in the interests of safety. They must also understand that intentional breaches of safety that jeopardize operations will not be tolerated. A positive safety culture is essential for the effective operation of an SMS. However, the culture of an organization is also shaped by the existence of a formal SMS. An organization should therefore not wait until it has achieved an ideal safety culture before introducing an SMS. The culture will develop as exposure and experience with safety management increases. Figure 11.5 shows common attributes of a positive safety culture.

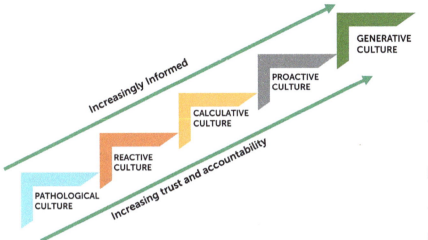

Figure 11.5  Common Attributes of a Positive Safety Culture

## Indications of Positive Safety Culture

- Senior management places strong emphasis on safety as part of the strategy of controlling risks and minimizing losses.
- Decision-makers and operational personnel hold realistic views of the short and long-term hazards involved in the organization's activities.
- Management fosters a climate in which there is a positive attitude toward criticisms, comments and feedback from lower levels of the organization on safety matters.
- Management does not use their influence to force their views on subordinates.
- Management implements measurestominimizetheconsequencesofidentifiedsafetydeficiencies.

Safety must not only be recognized but promoted by the senior management team as the organization's primary core value. Procedures, practices, training and the allocation of resources clearly demonstrate management's commitment to safety.

The key elements of promoting safety within any organization are:

- Safety Culture – Support the expansion of a positive safety culture throughout the organization by:
  - Widely distributing and visibly posting organizational safety policy and mission statements signed by senior management.
  - Clearly communicating safety responsibilities for all personnel.

- ▪ Visibly demonstrating commitment to safety through everyday actions.
- ▪ Implementing a "Just Culture" process that ensures fairness and open reporting in dealing with human error.
- ▪ Safety Education
- ▪ Widely communicated status on safety performance related to goals and targets
- ▪ Communication of all identified safety hazards
- ▪ Overview of recent accidents and incidents
- ▪ Communication of lessons learned that promote improvement in SMS
- ▪ Safety Training
- ▪ Initial "new employee" safety training
- ▪ Recurrent safety training for all employees
- ▪ Document, review and update training requirements
- ▪ Define competency requirements for individuals in key positions
- ▪ Introduce and review safety policies
- ▪ Review of safety reporting processes
- ▪ Safety Communication
- ▪ Communicate the realized benefits of SMS to all employees
- ▪ Implement a safety feedback system with appropriate levels of confidentiality that promote participation by all personnel in the identification of hazards
- ▪ Communicate safety information with employees through:
- ▪ Safety newsletters
- ▪ Bulletin board postings
- ▪ Safety investigation reports
- ▪ Internet website

## Chapter Questions

1. What are the four components, or pillars, of SMS?
2. What is the difference between the Accountable Executive and Key Safety Personnel?
3. Explain Safety Culture.

All information in this chapter is consistent with the information and guidance contained in other documents including:

- ICAO Doc 9859 Safety Management Manual
- FAA SMS Framework, SMS Assurance Guide and SMS Implementation Guide, as revised (these documents are the nucleus of the FAA Advisory Circular (AC) 120-92A
- FAA 14 CFR Part 5
- FAA SMS Voluntary Program
- FAA SMS Framework & Assurance Guide – Rev. 2
- Transport Canada Safety Management Manual TP 13739

In previous chapters we have talked about accident investigation theory and the accident investigation process. Accident investigation is a necessary process to learn what has actually happened and to try and prevent the same thing from happening in the future. This is called reactive safety. It is reactive in the sense that the accident or incident has already occurred. But, what if we could learn about the issues or problems before they actually become an accident? What if we were proactive about safety, instead of reactive? We might be able to actually save lives by preventing the accident from ever happening.

That is the idea behind proactive flight safety programs. These programs are an integral part of Safety Management System (SMS) and are the primary data sources for an SMS program. In this chapter, the programs we will talk about will be the Flight Operational Quality Assurance (FOQA) program, the Aviation Safety Action Program (ASAP), and the Line Operations Safety Audit (LOSA) program. These programs are all designed to measure and evaluate different aspects of the safety of the flight operations in an organization. This includes the overall effectiveness of the safety system itself.

## Flight Operational Quality Assurance (FOQA)

The Flight Operational Quality Assurance (FOQA) program is a safety program that uses data from a recording device on the aircraft itself. This is also known as Flight Data Monitoring (FDM). In this book, we will refer to these programs as FOQA programs. The recorded data can come from a few different sources on the aircraft. The conventional sources used might be the flight data recorder (FDR) or a quick access recorder (QAR). The QAR is a separate recorder that is made for recording aircraft data such as that used for FOQA. The retrieval of the data depends on the media type used by that particular QAR. The most common media types are an optical disk or a PC card. Maintenance personnel either send the media device to the safety office for downloading or it is downloaded at a workstation in the maintenance area itself remove the media at a certain interval. Wireless retrieval of data is becoming more common as more operators upgrade recording devices on the aircraft or upgrade the actual aircraft themselves. Most modern aircraft used by corporate operators or airlines come with a data recording device. Often this device has no physical media to remove, as data retrieval is completely wireless.

Figure 12.1 below shows the data retrieval process.

So, what kind of data is being recorded and downloaded? This data usually includes common parameters such as; airspeed, altitude, aileron or rudder position, landing gear down/up, etc. Older or less sophisticated aircraft may measure 50 or 100 parameters. The newest technology aircraft such as the Boeing 787 or the Airbus 380 can measure upwards of 3,500 parameters. The more parameters an aircraft records, the more data you can gather, and the better you will be able to really tell how safe or efficient your operation really is.

The more data coming off an aircraft, the tougher it can be to decipher what it is trying to tell you. This is where an operator will use computer software to help analyze the data. There are software systems available today that will look for certain "events" that an operator will set the system to look for. These events can be anything from and unstable approach to a high bank angle. By flagging these events, it tells a member of the FOQA team where in the data to look in order to analyze the issues that might be occurring in the operation.

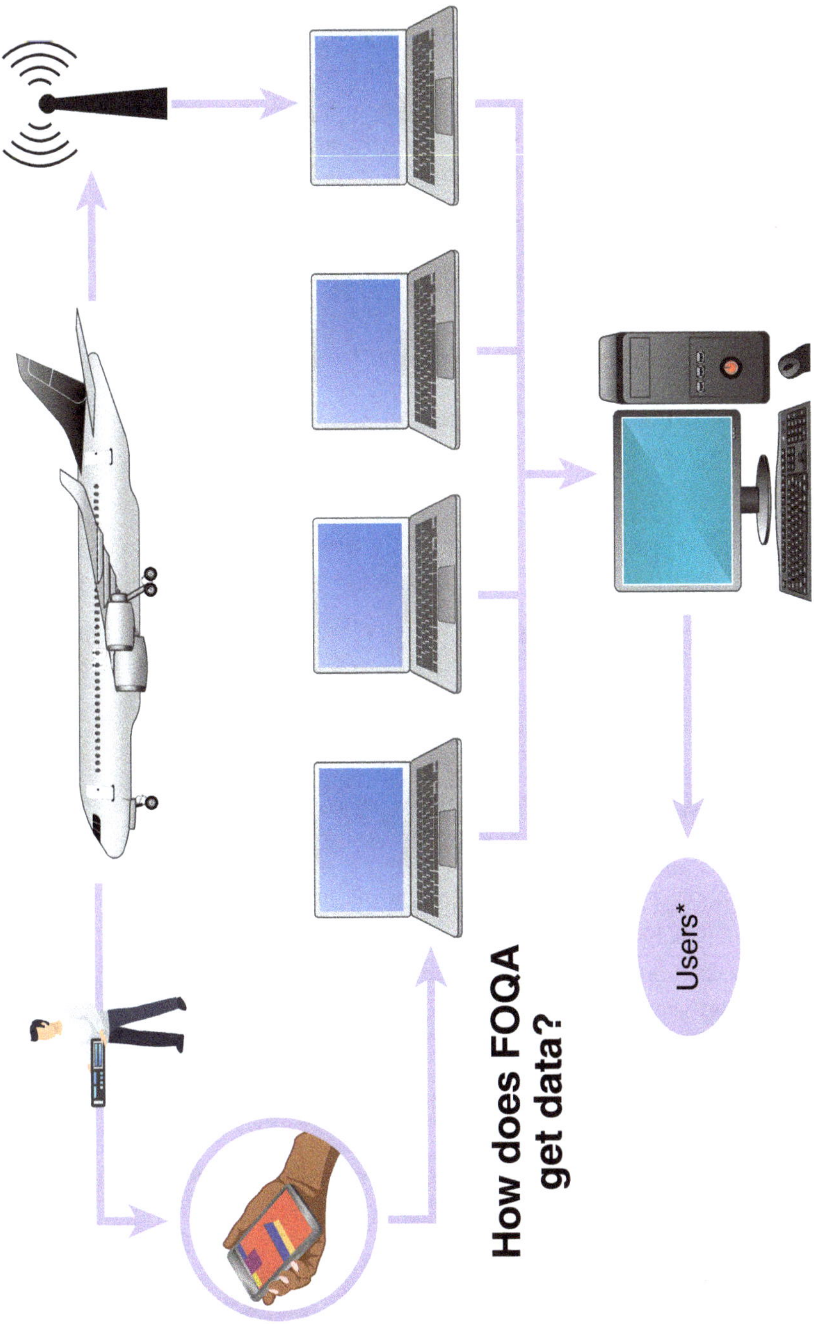

**How does FOQA get data?**

Users*

**Figure 12.1** Data Retrieval Process

© Kendall Hunt Publishing Company

Most FOQA programs, especially in the U.S., also de-identify the information and data as it comes into the analyses software. This de-identification takes the flight number and date of any particular flight and puts it in a separate file, accessible only by a designated "gatekeeper." This gatekeeper is usually a pilot who might represent the pilot's union, or association. If a particular event or flight that is being analyzed or has been flagged by the software system for review needs more information, the gatekeeper can use the identified information to contact the crew involved.

In terms of regulatory oversight, it depends on the regulating body. In the U.S. the FAA recommends that operators have a FOQA program through Advisory Circular 120-82. Specifically, it says:

*"... Provides guidance on one means, but not necessarily the only means, of developing, implementing, and operating a voluntary Flight Operational Quality Assurance (FOQA) program that is acceptable to the Federal Aviation Administration (FAA)."*

On the other hand, ICAO requires that certain operators have a "flight data programme" through this language:

*3.6.3: From 1 January 2005, an operator of an aeroplane of maximum certificated take-off weight in excess of 27,000kg shall establish and maintain a flight data analysis programme as part of its accident prevention and flight safety programme.*

*3.6.4: A flight data analysis programme shall be non-punitive and contain adequate safeguards to protect the source(s) of the data.*

The future of FOQA will undoubtedly bring more technological improvements, such as better wireless data transmission and more powerful FOQA analysis software. Research is also being supported and funded by the FAA to create better ways of analyzing data across the whole aviation system, as well as bringing the FOQA program to smaller general aviation aircraft and operators.

## Aviation Safety Action Programs (ASAP)

The ASAP program is an employee reporting system where in exchange for some sort of immunity from company (or FAA) discipline, the employee will report important safety information. This safety information may otherwise never be known or obtained. Most ASAP programs claim that most of the safety information reported by employees is sole source. Pilot reporting ASAP programs are the most popular in aviation organizations, but many operators also have ASAP programs for their mechanics, dispatchers, load planners, or flight attendants. The FAA even has an ASAP program for their air traffic controllers called the Air Traffic Safety Action Program (ATSAP). The FAA has been recently working with specific airlines to work on a program where data from a specific reported event has been matched up with the same event (if reported) by an airline. This program is called fusion. There are plans to add other data sources, such as FOQA, later.

So, how does ASAP actually work? As stated earlier, an agreement between the employee representatives (usually an employee union or organization), the employer, and the FAA allow for immunity of company discipline or FAA certificate action in exchange for the reported information. Once the report is submitted, the report is read and categorized by someone in the operator's safety department. The person analyzing the information will then forward the report to anyone they think may need the information. An example of this might be a report about a maintenance issue being forwarded to Technical Operations.

Most operators will also have some sort of event review committee (ERC) that will also review the submitted report. This committee usually consists of a representative of the employee association, the employer (operator), and the FAA. This committee meets on a regular basis to review the events and decide if more information may be needed from the employee(s). They can also decide that the employees involved may need more training. In the case of a flight crew, this training may be over the phone or even a session with an instructor in an aircraft simulator. In the case of an FAA approved ASAP program, the committee also has the right to decide if a report will actually be included in the program, in order to gain the protections to the employee listed earlier. A report might be exclude from the ASAP program and those protections due to several issues. These issues can range from illegal substances (drugs) being involved to an intentional disregard for safety by the employees who submitted the report. Figure 12.2 shows a typical ASAP process.

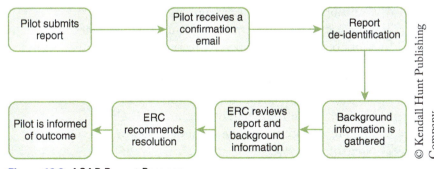

**Figure 12.2** ASAP Report Process

## Line Operated Safety Observation (LOSA)

LOSA is a safety program where qualified observers (usually line pilots), sit in a jumpseat on the flight deck during a flight. The job of these observers is to watch and see how normal operations are being flown. The number of observers used, as well as the number of flights observed, varies based on the size of an operation. Once the observations have been completed, the observation data is compiled, usually by an outside organization, as they will be unbiased. The data is analyzed to see what is actually happening on the flight deck in normal operations. An example of an issue that might be observed on flights might be a missed checklist item. If this item is missed on many flights observed, then a change may need to be made to the checklist to make the item more prominent.

A common complaint about LOSA observations is that the flight crews may not act in a normal manner due to being observed. Even though an observer usually has to ask for permission to observe the flight as well as letting the crew know that the observation is a non-jeopardy event for the crew, there is no way to hide that they are still being observed. So, the question remains; are you getting observations of what is really happening on the flight deck?

All three of the flight safety programs described can be great assets to an operation as long as the data is used correctly, and for the purposes of safety. Individually they are all valuable programs but when used in conjunction with each other, they are at their best. Successful SMS programs will include one or more of these programs, in order to get as much safety data into their operation. These proactive safety programs have been proven to decrease accidents. As technology improves, these programs will only get better and help save lives in aviation.

## Chapter Questions

1. FOQA, ASAP, and LOSA programs are all designed to:
   A) Contribute to an SMS program
   B) Enhance an operator's safety by being proactive instead of reactive
   C) Work better when an operator has all three programs
   D) All of the above
2. Flight Operational Quality Assurance programs
   A) Rely on submitted reports information by an employee
   B) Use recorded information off the aircraft itself
   C) Require an observer on the flight deck auditing the operations
   D) None of the above

3. FOQA programs are required by the FAA in the United States
   A) True
   B) False
4. All of the following are parties in an ASAP ERC, except:
   A) FAA
   B) Employee Group
   C) NTSB
   D) Operator

# Anti-Drug Programs in Aviation

**13**

## LEARNING OBJECTIVES

1. Appreciate the requirement for an Anti-Drug Plan.
2. Understand why the DOT mandates testing for illegal drugs.
3. Know who requires drug testing.
4. Know who must submit to a test.
5. Identify the country that consumes the most illegal drugs.
6. Know the type of drugs that the DOT/FAA is testing for.
7. Know the definition of "Refusal to take a drug test."
8. Become familiar with shy bladder procedures.

## Background

In May 2009, the Department of Transportation (DOT) and the Federal Aviation Administration (FAA) published the Drug and Alcohol Testing Program rule. That rule moved the drug and alcohol testing regulations into a new Title 14 of the Code of Federal Regulations (CFR) Part 120. Title 14 of the Code of Federal Regulations (CFR) Part 120 requires the establishment of a drug and alcohol testing program designed to prevent accidents and injuries that result from the use of prohibited drugs and the misuse of alcohol.

The transportation industry drug and alcohol testing program is a critical element of the DOT's safety mission. Pilots, air traffic controllers, aircraft mechanics, dispatchers, truck drivers, subway operators, ship captains, pipeline controllers, locomotive engineers, armed security personnel and bus drivers – among others – have an critical responsibility to the public and we cannot let their performance be compromised by drugs or alcohol.

**145**

Employers must also have strong drug and alcohol testing programs and employees must immediately be removed from safety-sensitive duties if they have violated drug and alcohol testing rules. Employees must not be returned to safety-sensitive duties until they have been referred for evaluation and have successfully complied with treatment recommendations.

## Requirements for an Anti Drug Program

The U.S. DOT has permitted some employers holding FAA certificates to implement their own anti-drug program. Some of the employers are 14 CFR part 121, part 135, part 141, air tour operators as defined under 14 CFR part 91, § 91.147, or air traffic control facilities not operated by the FAA or by or under contract to the U.S. military. These employers are still mandated to follow the FAA drug and alcohol testing program and still must comply with the following Code of Federal Regulations:

- Title 49 CFR part 40, Procedures for Transportation Workplace Drug and Alcohol Testing Programs, and
- Title 14 CFR part 120, Drug and Alcohol Testing Program

The testing procedures are established in the DOT's Procedures for Transportation Workplace Drug and Alcohol Testing Programs, Title 49 CFR Part 40. FAA-mandated drug and alcohol testing program must:

- Ensure that no one is hired for or transferred into a safety-sensitive function without first conducting a pre-employment drug test and receiving a verified negative test result. More information regarding pre-employment drug testing may be found under §120.109(a). Pre-employment alcohol testing is not required, however, may be implemented according to §120.217(a).
- Conduct a drug and alcohol records check, after obtaining an employee's written consent, requesting the information included in 49 CFR part 40, § 40.25(b) from DOT-regulated employers who have employed this individual during the two years prior to the date of application or transfer. Further requirements are explained in § 40.25. You must obtain and review this information prior to the first time the employee performs safety-sensitive duties. When hiring pilots, you must comply with the requirements of the Pilot Records Improvement Act (PRIA) and request records for the previous five years. For more information about PRIA, including a PRIA form, please review the FAA's PRIA Advisory Circular (AC 120-68F).

- Educate and train employees on the effects and consequences of drug abuse and alcohol misuse, as well as supervisors who will make determinations of whether reasonable cause/suspicion testing is necessary.
- Ensure that employees are included in your random drug and alcohol testing pool and have an equal chance of being tested each time selections are made. You must conduct annual random testing at a minimum rate of 25% for drugs and 10% for alcohol. More information regarding random testing may be found under §120.109(b) and §120.217(c).

The FAA's web site has a Designated Employer Representative (DER) awareness page that includes a video series, compliance brochure, and several posters to use in your facility or office.

We encourage you to subscribe to updates when information becomes available online.

## Safety-Sensitive Functions

All individuals performing safety-sensitive functions directly or by contract (including subcontract at any tier) are subject to testing. Safety-sensitive functions (as described in §120.105 and §120.215) include:

- flight crewmember duties,
- flight attendant duties,
- flight instruction duties,
- aircraft dispatcher duties,
- aircraft maintenance and preventive maintenance duties,
- ground security coordinator duties,
- aviation screening duties,
- air traffic control duties, and
- operations control specialist duties.

## When will employees be tested?

The individuals who are performing these safety-sensitive functions must be subject to the following types of drug and alcohol testing (as described in §120.109 and §120.217):

- pre-employment,
- reasonable cause/suspicion,
- random,
- post-accident,
- return-to-duty, and
- follow-up.

## What conduct is prohibited by the DOT regulations?

**As a safety-sensitive employee:**

- You must not use or possess alcohol or any illicit drug while assigned to perform safety-sensitive functions or actually performing safety-sensitive functions.
- You must not report for service, or remain on duty if you...
  - Are under the influence or impaired by alcohol;
  - Have a blood alcohol concentration .04 or greater; (with a blood alcohol concentration of .02 to .039, some regulations do not permit you to continue working until your next regularly scheduled duty period);
  - Have used any illicit drug.
- You must not use alcohol within four hours (8 hours for flight crew members and flight attendants) of reporting for service or after receiving notice to report. You must not report for duty or remain on duty when using any controlled substance unless used pursuant to the instructions of an authorized medical practitioner.
- You must not refuse to submit to any test for alcohol or controlled substances.
- You must not refuse to submit to any test by adulterating or substituting your specimen.

## What country consumes the most illegal drugs?

Despite tough anti-drug laws, a new survey shows the U.S. has the highest level of illegal drug use in the world. The World Health Organization's survey of legal and illegal drug use in 17 countries, including the Netherlands and other countries with less stringent drug laws, shows Americans report the highest level of cocaine and marijuana use. For example, Americans were four times more likely to report using cocaine in their lifetime than the next closest country, New Zealand (16% vs. 4%), Marijuana use was more widely reported worldwide, and the U.S. also had the highest rate of use at 42.4% compared with 41.9% of New Zealanders. In contrast, in the Netherlands, which has more liberal drug policies than the U.S., only 1.9% of people reported cocaine use and 19.8% reported marijuana use.

## What type of drugs is the DOT/FAA testing for?

- Marijuana metabolites/THC
- Cocaine metabolites
- Amphetamines (including methamphetamine, MDMA)

- Opiates (including codeine, heroin (6-AM), morphine)
- Phencyclidine (PCP)
- Alcohol

## Can I use prescribed medications & over-the-counter (OTC) drugs and perform safety-sensitive functions?

Prescription medicine and OTC drugs may be allowed. The FRA requires that if you are being treated by more than one medical practitioner, you must show that at least one of the treating medical practitioners has been informed of all prescribed and authorized medications and has determined that the use of the medications is consistent with the safe performance of your duties. However, you must meet the following minimum standards:

- The medicine is prescribed to you by a licensed physician, such as your personal doctor.
- The treating/prescribing physician has made a good faith judgment that the use of the substance at the prescribed or authorized dosage level is consistent with the safe performance of your duties.

Best Practice: To assist your doctor in prescribing the best possible treatment, consider providing your physician with a detailed description of your job. A title alone may not be sufficient. Many employers give employees a written, detailed description of their job functions to provide their doctors at the time of the exam.

- The substance is used at the dosage prescribed or authorized. While a minority of states allow medical use of marijuana, federal laws and policy do not recognize any legitimate medical use of marijuana. Even if marijuana is legally prescribed in a state, DOT regulations treat its use as the same as the use of any other illicit drug.
- If you are being treated by more than one physician, you must show that at least one of the treating doctors has been informed of all prescribed and authorized medications and has determined that the use of the medications is consistent with the safe performance of your duties.
- Taking the prescription medication and performing your DOT safety-sensitive functions is not prohibited by agency drug & alcohol regulations. However, other DOT agency regulations may have prohibitive provisions, such as medical certifications.

Remember: Someagencieshaveregulationsprohibitinguseofspecificprescriptiondrugs, e.g. methadone, etc.... If you are using prescription or over-the-counter medication, check first with a physician, but do not forget to consult your industry-specific regulations before deciding to perform safety-sensitive tasks. Also be sure to refer to your company's policy regarding prescription drugs.

## Shy Bladder Procedures

If you fail to provide a sufficient amount of urine when directed, and it has been determined, through a required medical evaluation, that there was no adequate medical explanation for the failure, it may be considered a refusal to test.

When you report for testing, you must make an attempt to provide a specimen. If you are unable to provide a sufficient amount of urine for a drug test, the collector must begin the "shy bladder" procedures. This procedure requires that you remain at the collection site. The collector must urge you to drink 40 ounces of fluid over a three-hour period until you provide a sufficient amount of urine or the three- hour time period has elapsed.

After three hours, if you did not provide a sufficient amount of urine, the collector must contact your employer's Designated Employer Representative (DER) to advise him/her that you were not able to provide a specimen.

After the DER consults with the company's Medical Review Officer (MRO), you will be directed to obtain a medical evaluation within five days by a physician (or the MRO) who has expertise in the medical issues raised by your failure to provide a sufficient specimen. If the MRO makes a determination that a medical condition precludes you from providing a sufficient amount of urine, the test may be cancelled, and it would not be considered a refusal. However, if the MRO determines there is no medical condition that precludes you from providing a sufficient urine sample, the collection will be deemed a refusal to submit.

The Department of Transportation's Web site (http://www.dot.gov/odapc) includes an employee guide that may help you better understand your responsibilities for testing.

## What constitutes a post-accident test? What is the definition of an accident?

The Federal Aviation Administration's (FAA's) drug and alcohol testing regulation (14 CFR part 120) describes when an employer is required to conduct and when an employee must submit to post-accident drug and/or alcohol testing.

As soon as practicable following an accident, each employer must test each surviving safety-sensitive employee for the presence of marijuana, cocaine, opiates, phencyclidine (PCP), and amphetamines, or a metabolite of those drugs in the employee's system, and for alcohol, if that employee's performance either contributed to the accident or cannot be completely discounted as a contributing factor to the accident.

For post-accident drug testing, the employee must be tested as soon as possible but not later than 32 hours after the accident.

For post-accident alcohol testing, the employee must be tested as soon as possible but the time of testing cannot exceed 8 hours from the time of the accident. If a test is not administered within 2 hours following the accident, the employer must prepare and maintain on file a record stating the reasons why the test was not promptly administered. If a test is not administered within 8 hours following the accident, the employer must cease attempts to administer an alcohol test and must prepare and maintain the same record.

The decision not to administer a test must be based on the employer's determination, using the best information available at the time of the determination that the employee's performance could not have contributed to the accident.

The FAA and the National Transportation Safety Board (NTSB) define an *accident* as an occurrence associated with the operation of an aircraft which takes place between the time any person boards the aircraft with the intention of flight and all such persons have disembarked, AND in which any person suffers death or **serious injury** or in which the aircraft receives **substantial damage**. The NTSB regulations (49 CFR part 830) define "serious injury" and "substantial damage" as follows:

"**Serious injury** means any injury which: (1) Requires hospitalization for more than 48 hours, commencing within 7 days from the date of the injury was received; (2) results in a fracture of any bone (except simple fractures of fingers, toes, or nose); (3) causes severe hemorrhages, nerve, muscle, or tendon damage;

involves any internal organ; or (5) involves second- or third-degree burns, or any burns affecting more than 5 percent of the body surface."

"**Substantial damage** means damage or failure which adversely affects the structural strength, performance, or flight characteristics of the aircraft, and which would normally require major repair or replacement of the affected component. Engine failure or damage limited to an engine if only one engine fails or is damaged, bent fairings or cowling, dented skin, small punctured holes in the skin or fabric, ground damage to rotor or propeller blades, and damage to landing gear, wheels, tires, flaps, engine accessories, brakes, or wingtips are not considered "substantial damage" for the purpose of this part."

Monetary damage is not a factor in determining what constitutes an "accident."

## Should I refuse a test if I believe I was unfairly selected for testing?

Rule of Thumb: Comply then make a timely complaint. If you are instructed to submit to a DOT drug or alcohol test and you don't agree with the reason or rationale for the test, take the test anyway. Don't interfere with the testing process or refuse the test. After the test, express your concerns to your employer through a letter to your company's dispute resolution office, by following an agreed upon labor grievance or other company procedures. You can also express your concerns to the appropriate DOT agency drug & alcohol program office. (See contact numbers listed in the Appendix.) Whomever you decide to contact, please contact them as soon as possible after the test.

## What is considered a refusal to test?

DOT regulations prohibit you from refusing a test. The following are some examples of conduct that the regulations define as refusing a test (See 49 CFR Part 40 Subpart I & Subpart N):

- Failure to appear for any test after being directed to do so by your employer.
- Failure to remain at the testing site until the testing process is complete.
- Failure to provide a urine or breath sample for any test required by federal regulations.
- Failure to permit the observation or monitoring of you providing a urine sample. (Please note tests conducted under direct observation or monitoring occur in limited situations. The majority of specimens are provided in private.)
- Failure to provide a sufficient urine or breath sample when directed, and it has been determined, through a required medical evaluation, that there was not adequate medical explanation for the failure.
- Failure to take a second test when directed to do so.
- Failure to cooperate with any part of the testing process.
- Failure to undergo a medical evaluation as part of "shy bladder" or "shy lung" procedures.
- Failure to sign the blood/urine submission form.
- Providing a specimen that is verified as adulterated or substituted.

- Failure to cooperate with any part of the testing process (e.g., refuse to empty pockets when directed by the collector, behave in a confrontational way that disrupts the collection process, fail to wash hands after being directed to do so by the collector).
- Failure to follow the observer's instructions [during a direct observation collection] to raise your clothing above the waist, lower clothing and underpants, and to turn around to permit the observer to determine if you have any type of prosthetic or other device that could be used to interfere with the collection process.
- Possess or wear a prosthetic or other device that could be used to interfere with the collection process.
- Admit to the collector or MRO that you adulterated or substituted the specimen.

## DOT Publication: "What Employees Need to Know About DOT Drug & Alcohol Testing"

This publication was produced by the U.S. Department of Transportation (DOT) to assist safety-sensitive employees subject to workplace drug & alcohol testing in understanding the requirements of 49 CFR Part 40 and certain DOT agency regulations. Nothing in this publication is intended to supplement, alter or serve as an official interpretation of 49 CFR Part 40 or DOT agency regulations. This publication is for educational purposes only and can be downloaded at http://www.transportation.gov/odapc

## Chapter Questions

1. What country consumes the most illegal drugs?
2. Explain the term shy bladder procedures.
3. What is considered refusal to take a test?

# General Aviation Accidents

## What is General Aviation?

General aviation (GA) is the umbrella term for any operation that does not operate under Parts 121, 135, or 129. In 2010, general aviation accounted for 96 percent of all aviation accidents, 97 percent of fatal aviation accidents, and 96 percent of all fatalities for U.S. civil aviation. In addition, general aviation accounted for 51 percent of the estimated total flight time of all U.S. civil aviation in 2010. Figure 14.1 shows the most recent analysis of total and fatal general aviation accidents from 2001 through 2010. Figure 14.2 shows the number of total and fatal accident aircraft for the same period.

The number of general aviation accidents declined over the decade, however the number of fatal accidents remained relatively stable over the ten-year period. Figure 14.3 shows the estimated total flight hours for general aviation based on the GA Survey. Figure 14.4 plots the total and fatal accident rates for general aviation.

General aviation covers a wide range of operations and aircraft, from powered parachutes and light sport aircraft to turboprops and jets used for a variety of flying. General aviation includes some types of commercial activities. The sections that follow discuss the top five types of general aviation operations based on their number of accidents and, where appropriate,

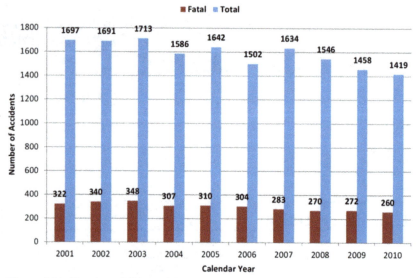

**Figure 14.1** Number of GA Accidents and Fatal Accidents
*Source:* NTSB

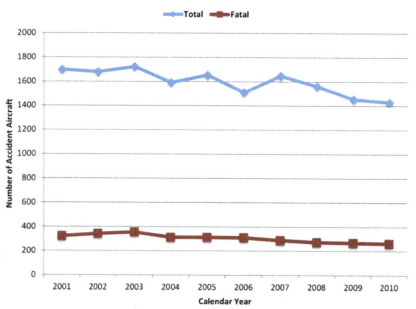

**Figure 14.2** Number of Aircraft Involved in GA Accidents
*Source:* NTSB

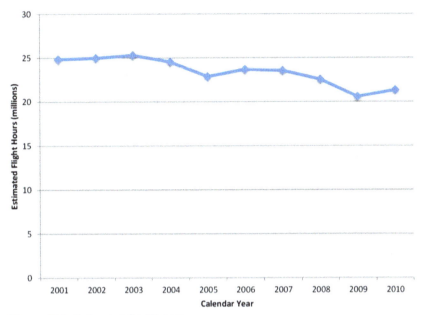

**Figure 14.3** Estimated GA Flight Hours

*Source:* NTSB

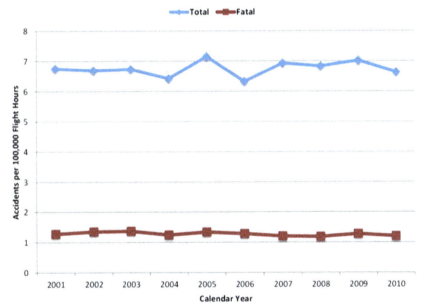

**Figure 14.4** Total and Fatal Accident Rates for GA

*Source:* NTSB

examine the types of aircraft involved in these accidents. Table 14.1 shows the purpose of flight, by aircraft type, for the 1,433 aircraft involved in general aviation accidents in 2010.

The majority of general aviation accidents in 2010 involved personal flying in fixed-wing airplanes, which accounted for 64 percent (912) of the accidents, followed by flight instruction in fixed-wing airplanes, which accounted for 10 percent (140) of the accidents. Aerial application, business, and public aircraft operations followed next in accident frequency in 2010. Fixed-wing airplanes accounted for 87 percent (1,242) of all general aviation accidents, helicopters accounted for 8 percent (120), and all other aircraft accounted for the remaining 5 percent (71).

## Personal Flying

From local currency flights to long distance, cross-country flights, personal flying involves a wide variety of flight activities and aircraft. Most personal

**Table 14.1**  Number of GA Accident Aircraft by Aircraft Type and Purpose of Flight

| Purpose of Flight | Fixed Wing | Helicopter | Balloon | Glider | Other Aircraft | Total |
|---|---|---|---|---|---|---|
| Personal | 912 | 33 | 5 | 25 | 24 | 999 |
| Flight Instruction | 140 | 23 | 1 | 4 | 4 | 172 |
| Aerial Application | 60 | 23 | 0 | 0 | 0 | 83 |
| Business | 38 | 7 | 1 | 0 | 0 | 46 |
| Public Aircraft Operations | 15 | 8 | 0 | 0 | 0 | 23 |
| Positioning | 16 | 6 | 0 | 0 | 0 | 22 |
| Other Work Use | 10 | 7 | 0 | 1 | 0 | 18 |
| Flight test | 12 | 2 | 0 | 0 | 0 | 14 |
| Ferry | 11 | 0 | 0 | 0 | 0 | 11 |
| Aerial Observation | 2 | 5 | 0 | 0 | 0 | 10 |
| External Load | 0 | 5 | 0 | 1 | 0 | 6 |
| Unknown | 6 | 1 | 0 | 0 | 0 | 7 |
| Air Race/Show | 1 | 0 | 2 | 2 | 0 | 5 |
| Banner Towing | 5 | 0 | 0 | 0 | 0 | 5 |
| Executive/ Corporate | 5 | 0 | 0 | 0 | 0 | 5 |
| Glider Tow | 3 | 0 | 0 | 1 | 0 | 4 |
| Skydiving | 3 | 0 | 0 | 0 | 0 | 3 |
| Total | 1,242 | 120 | 9 | 34 | 28 | 1,433 |

*Source:* NTSB

flying was conducted in fixed-wing airplanes, leading to a higher exposure for these types of aircraft. In particular, the GA Survey estimates that 78 percent of personal flying was conducted in single-engine, piston-driven airplanes in 2010. The volume of personal flying decreased significantly between 2003 and 2010.

Figure 14.5 shows the number of total and fatal personal flying accidents in 2010. Over the ten-year period, personal flying accidents ranged from 1,080 in 2001 to 990 in 2010.

Figure 14.6 plots the accident rates associated with these data. The accident rate for personal flying increased over the decade, rising to about 12 accidents per 100,000 hours. The fatal accident rate for the 10-year period remained stable.

Figure 14.7 shows the defining events for the 999 aircraft involved in personal flying accidents for the last year of analysis.

System and component malfunction or failure accounted for about 20 percent of the non-fatal and about 13 percent of the fatal accidents in 2010. Loss of control in flight accounted for the largest proportion of fatal accidents (about 38 percent). Figure 14.8 shows the phase of flight corresponding to each defining event for personal flying accidents.

The majority of total personal flying accidents occurred during the landing phase, followed closely by the enroute phase. The majority of fatal personal flying accidents occurred during the maneuvering phase.

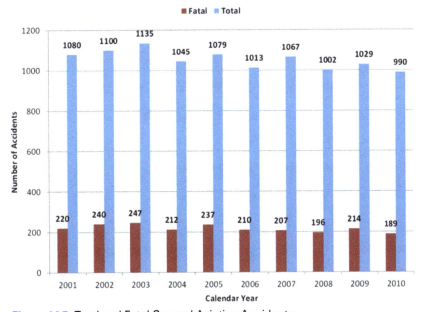

**Figure 14.5** Total and Fatal General Aviation Accidents
*Source:* NTSB

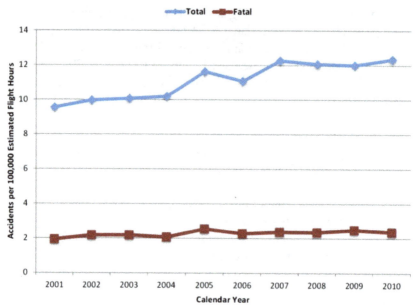

**Figure 14.6** Accident Rates for Personal Flying

*Source:* NTSB

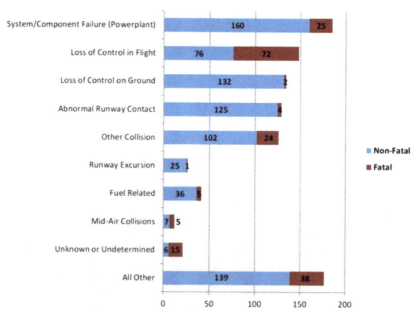

**Figure 14.7** Defining Events for Personal Flying Accidents

*Source:* NTSB

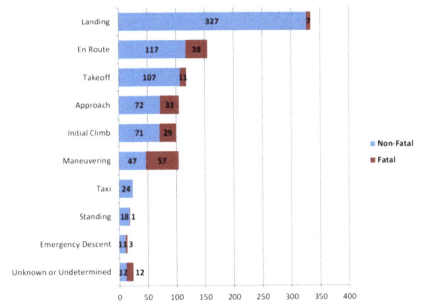

**Figure 14.8** Phase of Flight for Personal Flying Accidents
*Source:* NTSB

Pilots involved in personal flying accidents had an average total flight time of 2,863 total hours, with   a range of 20 to 31,270 hours. The average time in the type of accident aircraft was 460 hours, with a range of 1 to 10,000 hours.

## Preliminary Analysis for 2010–2014

While the number of fatal general aviation accidents over the last decade has decreased, so have the estimated of total GA flight hours, likely due to economic factors.

The general aviation fatal accident rate appears to have remained relatively static based on the FAA's flight hours estimates. Table 14.2 shows the FAA's preliminary estimate for FY 2014 is a fatal accident rate of 1.09 with 251 GA fatal accidents with 434 fatalities. In 2013, the fatal accident rate was 1.11 fatal accidents per 100,000 hours, with 449 GA fatal accidents. In 2012, the fatal accident rate was 1.09 fatal accidents per 100,000 hours flown, with 267 GA fatal accidents. In 2011, the fatal accident rate was 1.12 fatal accidents per 100,000 hours flown, with 469 GA fatalities. In 2010, the fatal accident rate was 1.10 fatal accidents per 100,000 hours flown, with 272 GA fatal accidents.

**Table 14.2**  Previous 5-year GA fatal accident rates and numbers

| Year | GA Fatal Accidents per 100,000 Hours | GA Fatal Accidents | GA Fatalities |
|------|--------------------------------------|--------------------|---------------|
| FY10 | 1.10 | 272 | 471 |
| FY11 | 1.12 | 278 | 469 |
| FY12 | 1.09 | 267 | 442 |
| FY13 | 1.11 | 259 | 449 |
| FY14 (est) | 1.09 | 251 | 434 |

*Source:* FAA

## The Top 10 Leading Causes of Fatal General Aviation Accidents 2001–2013:

1. Loss of Control Inflight
2. Controlled Flight into Terrain
3. System Component Failure – Powerplant
4. Low Altitude Operations
5. Other
6. System Component Failure – Non-Powerplant
7. Fuel Related
8. Unknown or Undetermined
9. Windshear or Thunderstorm
10. Midair Collisions

## NTSB Most Wanted Safety Improvements

### What is the issue?

While commercial aviation continues to have a strong safety record of 2 years without a fatal accident, the NTSB continues to investigate about 1,500 accidents each year in general aviation. In many cases, pilots did not have the adequate knowledge, skills, or recurrent training to fly safely, particularly in

**Figure 14.9**  NTSB Most Wanted Safery Improvement List

questionable weather conditions. In addition, the more sophisticated "glass" cockpit displays present a new layer of complications for general aviation pilots. In addition, not only are pilots dying due to human error and inadequate training, but also they are frequently transporting their families who suffer the same tragic fate.

## What can be done

In our general aviation accident investigations, the NTSB sees similar accident circumstances time after time. Adequate education, training, and screening for risky behavior are critical to improving general aviation safety. For example, guidance materials should include information on the use of Internet, satellite, and other data sources for obtaining weather information. Training materials should include elements on electronic primary flight displays, and pilots should have access to flight simulators that provide equipment-specific electronic avionics displays. Knowledge tests and flight reviews should test for awareness of weather, Use of instruments, and use of "glass" cockpits. In addition, there should be a mechanism for identifying at-risk pilots and addressing risks so that both the pilot and passengers can safely fly. Human error in general aviation accidents is not solely a pilot problem. Aircraft maintenance workers should also be required to undergo recurrent training to keep them up to date with the best practices for inspecting and maintaining electrical systems, circuit breakers, and aged wiring.

## Statistics

General aviation has the highest aviation accident rate within civil aviation. The rate is 6 times higher than for small commuter operators and 40 times higher than for transport category operations. Although the overall general aviation accident rate has remained relatively steady at an average of 6.8 per 100,000 flight hours, the components of that figure have changed dramatically over the last 10 years. In particular, personal flying accident rates have increased 20 percent, while the fatal accident rate has increased 25 percent over the same 10-year period. The NTSB sees this statistic play out frequently, having investigated an average of 1,500 general aviation accidents each year, in which more than 400 pilots and passengers are killed annually.

## Related Reports

### Safety Recommendation Letter October 12, 2005

*NTSB Recommendation Numbers A'()5-024 through A-05-029, adopted on October 12, 2005*

Aviation Accident Report: In flight Fire, Emergency Descent, and Crash in a Residential Area Cessna 31 OR, N501 N, Sanford, FL July 10, 2007

### NTSB Report Number: AAR-09-01, Adopted on January 28, 2009

Safety Study: Introduction of Glass Cockpit Avionics into Light Aircraft NTSB Report Number: SS-1 0/01, adopted on March 9, 2010.

## Example Accident #1

NTSB Identification: WPR14LA005 HISTORY OF FLIGHT

On October 4, 2013, at 1300 Mountain Standard Time, a Cessna 340A, N312GC, collided with a radio tower while maneuvering near Paulden, Arizona. The airplane was registered to and operated by the pilot under the provisions of 14 Code of Federal Regulations Part 91 as a personal flight. The private pilot and three passengers were fatally injured. The airplane was substantially damaged. Visual meteorological conditions prevailed at the time and no flight plan was filed. The airplane departed from Bullhead City, Arizona about 1130.

A summation of several witnesses at a gun club reported the airplane made one high speed, low pass from north to south over the club's buildings, and then maneuvered around for another low pass from east to west. During the second low pass over the buildings, the airplane collided with an approximate 50-foot-tall radio tower. The base of the tower was triangular shaped with each of the sides about two feet in length. About 10 feet of the tower was sheared off by the airplane's right wing. After the impact, the airplane rolled to the right almost inverted and subsequently impacted trees and terrain approximately 700 feet southwest of the initial impact point. One witness reported that the airplane remained in a straight and level attitude until the impact. This witness also stated that about three to four years prior to this accident, the pilot, a client of the gun club, had "buzzed" the club and had been told never to do so again.

### Personnel information

The 57-year-old pilot held a private pilot certificate and was rated in single and multi-engine land, helicopter and instrument airplane. Federal Aviation

Administration (FAA) records indicate that the pilot held a third-class medical certificate dated August 16, 2012. A limitation indicated that the pilot must have glasses available for near vision. At the time of the medical application, the pilot reported a total flight time of 4,006 hours.

## Aircraft information

The six seat, low wing, retractable landing gear airplane, serial number 340A0023, was manufactured in 1975. It was powered by two Continental Motors TSIO-520 engines.

Maintenance records indicated that the last annual inspection was completed on April 25, 2013. At that time the airframe total time was reported as 7,690.0 hours. The entries for both the left and right engine indicated a tachometer time of 2,491.0 hours and 609.5 hours since major overhaul.

## Communications

At 1128, the pilot contacted Lockheed Martin Flight Service, reported that he was on the ground at Bullhead City, and planned to fly to Prescott, Arizona. The pilot inquired about a temporary flight restriction (TFR) at Prescott and wanted to know what time the airport would reopen. The specialist confirmed that the airport would be open at 1130. Runways 12 and 30 were closed to fixed wing traffic due to air show activities. The specialist then inquired if the pilot wanted weather advisories. The pilot replied that he did and the specialist informed the pilot of an AIRMET for moderate turbulence below 14,000 feet all along the route of flight. The conversation ended at 1131.

At 1204, the pilot contacted flight watch reporting that he was 15 miles south of Kingman, Arizona, and wanted to confirm the ending time of the TFR at Prescott. The specialist confirmed that the TFR ended at 1130. Before the conversation ended the pilot provided a pilot report of the in-flight weather conditions. The pilot reported that he was at 11,500 feet and experiencing light chop to smooth air. The winds were from 340 degrees at 12 knots and the outside air temperature was 1°C. The conversation ended at 1209.

There were no further communications from the pilot. WRECKAGE AND IMPACT INFORMATION

The airplane's right wing collided with the approximate 50-foot tall radio tower shearing off the top 10 feet of the tower. The right wing folded up after the impact and the airplane began a roll to the right to the almost inverted position. The airplane subsequently collided with trees and terrain about 700 feet southwest of the initial impact point. A post-crash fire consumed the wreckage.

## Medical and pathological information

The office of the medical examiner, Yavapai County, performed an external examination on the pilot. The report indicated the cause of death as "blunt force trauma and thermal injury."

Toxicological samples were taken from the pilot and sent to the FAA Civil Aeromedical Institute in Oklahoma City, Oklahoma for analysis. The analysis revealed Metoprolol detected in urine and muscle; Rosuvastatin detected in urine and Zolpidem detected in urine and muscle.

## Tests and research

The wreckage was recovered to a secured facility in Phoenix, Arizona. A post-accident examination of the engines and the airframe revealed no evidence of a mechanical malfunction or failure that would have precluded normal operation. Detailed reports are in the public docket.

## Example Accident #2

Pilot/Race 177, The Galloping Ghost, North American P-51D, Reno, NV (11 fatal, 64 injured)

On September 16, 2011, an experimental, single-seat North American P-51D, N79111, collided with the airport ramp in the spectator box seating area following a loss of control during the National Championship Air Races unlimited class gold race at the Reno/Stead Airport (RTS), Reno, Nevada. The airplane was registered to Aero-Trans Corp (dba Leeward Aeronautical Sales), Ocala, Florida. The aircraft was operated by the commercial pilot as Race 177, The Galloping Ghost, under the provisions of Title 14 Code of Federal Regulations (CFR) Part 91. The pilot and 10 people on the ground sustained fatal injuries, and at least 64 people on the ground were injured (at least 16 of whom were reported to have sustained serious injuries). The airplane sustained substantial damage, fragmenting upon collision with the ramp. Visual meteorological conditions prevailed, and no flight plan had been filed for the local air race flight, which departed RTS about 10 minutes before the accident.

The NTSB determined that the probable cause of this accident was the reduced stiffness of the elevator trim tab system that allowed aerodynamic flutter to occur at racing speeds. The reduced stiffness was a result of deteriorated locknut inserts that allowed the trim tab attachment screws to become loose and to initiate fatigue cracking in one screw sometime before the accident flight. Aerodynamic flutter of the trim tabs resulted in a failure of the left trim tab link assembly, elevator movement, high flight loads, and a loss of control. Contributing to the accident were the

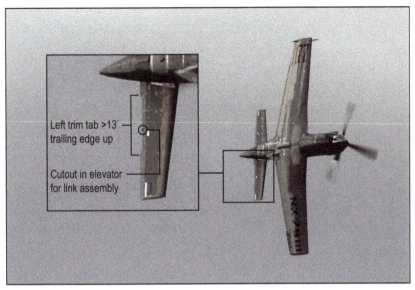

**Figure 14.10**  Reno Air Races, Pilot/Race 177, the Galloping Ghost, North American P-51D, Reno, Nevada, September 16, 2011

undocumented and untested major modifications to the airplane and the pilot's operation of the airplane in the unique air racing environment without adequate flight testing.

## Safety Recommendations Issued

As a result of its investigation, the NTSB issued four new safety recommendations to the Reno Air Racing Association, five new recommendations to the National Air-racing Group Unlimited Division, and one new recommendation to the FAA.

## Chapter Questions

1. How does the General Aviation accident rate compare to other aviation categories?
2. What phase of flight is the most hazardous?
3. Why has the NTSB put General Aviation Safety onto its Most Wanted Safety Improvement List?

# Air Traffic Control (ATC) Safety

**LEARNING OBJECTIVES**

1. Know the four components that makeup ATO SMS.
2. Know the three most crucial safety issues for the Federal Aviation Administration's (FAA).
3. Know the definition of a runway incursion.
4. Know types of runway incursions.
5. Know Technology available to make runways safer.
6. Know Best Practices for pilots and controllers.
7. Know Best Practices for airport personnel.

## Definitions

- Acceptance – The process whereby the event review committee (ERC) determines that a report meets VSRP requirements, receives it into the program for the relevant safety data contained therein, and provides the protective provisions of the program.
- ASAP – Aviation Safety Action Program. A defined partnership program where safety issues are resolved through corrective action rather than through punishment or discipline; the program includes collection, analysis, and retention of safety data reviewed and acted upon by an operating organization, their regulator, and the affected employee labor organization.
- ATSAP – Air Traffic Safety Action Program. The VSRP for air traffic control personnel based on the ASAP model as defined in this order and in the "FAA Air Traffic Organization (ATO) Air Traffic Safety

Contributed by Terra Jorgensen © Kendall Hunt Publishing Company

Action Program (ATSAP) for Air Traffic Personnel Memorandum of Understanding" signed by NATCA and the FAA.

- ATSAP report – A confidential written account of an event that involves an air traffic safety event or problem reported through ATSAP.
- Consensus – Unanimous or general agreement and the process to reach that status. Conceptually, consensus is related to cooperation—the process of working or acting together. For VSRPs, consensus refers to the voluntary agreement of all representatives of the ERC for a particular outcome.
- Corrective action request (CAR) – A CAR is a formal document identifying a nonconformance that is systemic in nature and requires a root cause analysis and modification.
- Credential action – Any action, including amendment and removal, taken by AOV toward an employees' AOV Credential (FAA Form 8060-66) as described in FAA Order 8000.90 and associated union negotiated agreements.
- Event review committee (ERC) – The group comprised of a representative from each party to a non-punitive safety-reporting program, which reviews and analyzes submitted confidential reports to identify actual or potential safety problems and ensure appropriate action is taken.
- Intentional falsification – As related to the exclusionary criteria, intentional falsification refers to knowingly misrepresenting facts with respect to required safety data.
- Mandatory occurrence report (MOR) – A report of certain types of safety events as defined by FAA Order JO 7210.632, Air Traffic Organization Occurrence Reporting.
- National Airspace System (NAS) systemic problem – Safety problem relating to the air traffic system as a whole. This would normally refer to a potential deficiency involving procedures, processes, training, culture, etc. that may be pervasive throughout the NAS.
- Non-sole source – Those reports that do not meet sole source criteria as defined below.
- Protective provision – The reporting incentives in this order and the applicable MOU that ensure a non-punitive environment for filing reports.
- Skill enhancement – Individually focused education and training designed to address an identified qualification issue of an employee in a skill or task.
- Safety check – An undocumented observation period requested by an employee, or required by an employee's manager. The objective of the observation is to confirm the employee's self-confidence in their ability to provide air traffic services after a serious safety event.

- Sole source – When all evidence of an event is discovered by, or otherwise based on a VSRP report or as otherwise designated by MOU or applicable chapters of this order.
- System corrective action – Those actions taken to correct identified deficiencies occurring beyond the individual. These could include issues pervasive throughout the system, or specific to the system itself.

## Introduction

The Federal Aviation Administration (FAA) provides air traffic services for the world's largest and busiest airspace. Tens of thousands of aircraft are guided safely and expeditiously every day through America's National Airspace System (NAS) to their destinations.

Every minute, every hour, every day, there are men and women at work to ensure the safety and efficiency of our national airspace system. According to the faa.gov website in 2016, Air Traffic Controllers handled 16,054,495 flights.

The NAS consists of 14,050 Air Traffic Controllers (ATC). ATC controls 24,101,568 square miles of the United States and Oceanic Airspace, of this 5,282,000 square miles is the United States domestic airspace. The airspace consists of 5,116 public airports, 14,485 private airports for a total of 19,601 US airports. There are 518 Air Traffic Control Towers, 24 Air Route Traffic Control Centers, 163 Terminal Radar Approach Controls and 236 Alaska Weather Cameras.

General Aviation, in 2015, contributed 24,142,000 flight hours. There were 164,200 fixed wing general aviation aircraft, 10,500 rotorcraft and 35,300 experimental/light sport aircraft.

Commercial operations consist of 6,676 commercial fleet aircraft. These aircraft make up a daily average of 26,527 scheduled passenger flights. There are 2,586,582 Domestic and International passengers every day. In 2016, freight operations were responsible for carrying 39.9 Billion pounds of freight.

Aviation contributes 5.1% to the U. S. gross domestic product (GDP). Aviation supports 10,600,000 jobs in the U.S. annually that creates $446.8 Billion in annual earnings.

Aviation Safety Program Management and Safety Management System (SMS) are discussed in Chapter 2 of this book. In 2008, the FAA Air Traffic Organization (ATO) implemented the Air Traffic Safety Action Program (ATSAP) for Air Traffic Controllers. It is paramount that the aviation industry maintains a safe and efficient NAS. This Chapter will discuss the ATO safety plan ATSAP, the top three safety concerns for the FAA and Best practices for ATC, Pilots and Airport Personnel.

## FAA ATO SMS

The Safety Management System (SMS) is a multidisciplinary, integrated, and closed-loop framework used to help maintain safe and efficient air navigation services and infrastructure throughout the NAS and United States–controlled international/oceanic airspace. The four components that make up the SMS are:

- Safety Policy. The requirements, standards, guidance, methods, and processes the Air Traffic Organization (ATO) uses to establish, execute, and improve the SMS, ensure NAS safety, and promote a positive safety culture.
- Safety Promotion. The communication of proper safety practices through advocacy of the principles of a positive safety culture; the conduct of employee training; compliance with ATO orders, policies, and guidance; and the use of data, processes, and tools to improve safety in daily ATO operations and in interactions with the NAS.
- Safety Assurance. The processes and procedures within the ATO SMS that ensure the ATO is operating according to expectations and requirements. Safety Assurance continually monitors ATO internal processes and operations to determine compliance with safety-related and SMS requirements and to ensure changes or deviations that may introduce risk to the NAS are addressed through the SRM process. Safety Assurance provides validation of SRM efforts for operations, systems, and equipment; identification of adverse safety trends through operational data collection and analysis; and the auditing of SMS performance, compliance, and processes.
- Safety Risk Management (SRM). The processes and procedures established and followed by ATO safety practitioners to identify hazards, analyze and assess their risks, determine safety performance targets, and implement and track appropriate risk controls for all air traffic operations, facilities, equipment, and systems in the NAS.

Figure 15.1 represents the relationship of the four SMS components in an integrated model. The integration and interaction of the four components is essential to managing the SMS effectively and fostering a positive safety culture.

## Safety Policy

The Air Traffic Organization (ATO) Safety Management System (SMS) is supported by numerous levels of policy and requirements, as depicted in Figure 15.2. These publications are explained below.

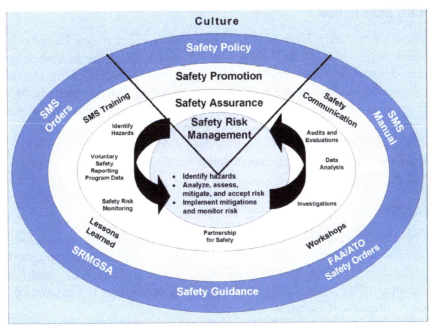

**Figure 15.1** The ATO Integrated Components of the SMS

*Source:* FAA.gov

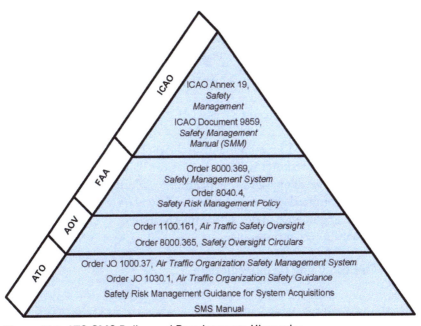

**Figure 15.1** ATO SMS Policy and Requirements Hierarchy

*Source:* FAA.gov

## ICAO SMS Policy

The FAA derives its high-level SMS policy from International Civil Aviation Organization (ICAO) policy. ICAO Annex 19, *Safety Management*, provides standards and recommended practices for safety management for member states and air traffic service providers. Additionally, ICAO Document 9859, *Safety Management Manual*, provides guidance for the development and implementation of the SMS for air traffic service providers. ICAO Document 9859 also provides guidance for safety programs in accordance with the international standards and recommended practices contained in Annex 19.

## FAA SMS Policy

The current version of FAA Order 8000.369, *Safety Management System*, describes the essential aspects of an SMS and provides implementation guidance to FAA organizations. This document is designed to create a minimum SMS standard that each FAA Line of Business (LOB) can follow to implement an SMS.

The current version of FAA Order 8040.4, *Safety Risk Management Policy*, provides risk management policy for FAA LOBs to follow when hazards, risks, and associated safety analyses affect multiple LOBs. The ATO must consider and, when necessary, use the provisions in this order when coordinating safety assessments with other FAA organizations. Safety and Technical Training (AJI) will function as the ATO liaison to interface with outside organizations. Within the ATO, AJI will adjudicate discrepancies among Service Units.

## AOV Order

The Air Traffic Safety Oversight Service (AOV) provides independent safety oversight of the ATO. FAA Order 1100.161, *Air Traffic Safety Oversight*, provides high-level SMS requirements of the ATO and AOV. When AOV involvement is required, AJI will function as the liaison between AOV and other ATO Service Units and organizations. Additional guidance from AOV will be submitted via Safety Oversight Circulars (SOCs) that provide information and guidance material that may be used by the ATO to develop and implement internal procedures.

## ATO SMS Policy and Requirements

FAA Order JO 1000.37, *Air Traffic Organization Safety Management System*, documents high-level SMS requirements, roles, and responsibilities. Additional requirements are contained within this SMS Manual. FAA

Order JO 1030.1, *Air Traffic Organization Safety Guidance*, establishes a method and a process for providing the ATO with supplemental guidance material pertinent to the SMS. The Safety Risk Management Guidance for System Acquisitions provides SMS requirements and guidance pertinent to programs proceeding through the FAA Acquisition Management System process. The ATO has also established Quality Assurance and Quality Control orders that govern safety data collection and the establishment of safety-related corrective actions. Those orders are as follows:

- FAA Order JO 7210.632, *Air Traffic Organization Occurrence Reporting*
- FAA Order JO 7210.633, *Air Traffic Organization Quality Assurance Program (QAP)*
- FAA Order JO 7210.634, *Air Traffic Organization (ATO) Quality Control*
- FAA Order JO 7200.20, *Voluntary Safety Reporting Program (VSRP)*

## Safety Promotion

One of the key functions of Safety Promotion is to provide communication channels between personnel on the operational front line and the appropriate safety organization. Safety Promotion is also furthered through programs such as workshops, lessons learned and SMS training.

The Voluntary Safety Programs Branch of the Flight Standards Services has been the FAA's leader in the development of safety programs designed to provide the FAA with safety information and instances of non-compliance that would otherwise not be known. The information provided through these dashboards, bulletins and corrective action requests, facilitates the identification of weaknesses in safety programs or procedures at individual, organizational, and systemic levels. The ability of the FAA to promote safe air transportation is greatly enhanced by the information gained through voluntary safety reporting programs.

## Safety Assurance

Safety Assurance provides validation of SRM efforts for operations, systems, and equipment; identification of adverse safety trends through operational data collection and analysis; and the auditing of SMS performance, compliance, and processes. The ATO maintains a positive safety culture using programs and initiatives listed below:

- **Recurrent Training:** Collaboratively-developed instruction for controllers, designed to maintain and update previously learned skills while promoting a positive safety culture.

- **Partnership for Safety:** A joint effort between the ATO and the National Air Traffic Controllers Association that encourages employees to become actively engaged in identifying local hazards and developing safety solutions before incidents occur.
- **Voluntary Safety Reporting Programs**
  - **Air Traffic Safety Action Program (ATSAP):** A confidential system for controllers and other employees to voluntarily identify and report safety and operational concerns.
  - **Confidential Information Share Program:** A program for the sharing and analysis of information collected through the ATSAP and airlines' Aviation Safety Action Programs to provide a more complete representation of the NAS.
  - **Technical Operations Safety Action Program (T-SAP):** A system for reporting safety-related events or issues pertaining to operations, equipment, personnel, or anything believed to affect safety in the NAS for technicians and other Technical Operations employees. The FAA created T-SAP in conjunction with ATSAP as controllers and technicians rely on one another to keep the NAS operating and safe.
  - **Lessons Learned:** Lessons learned are used to improve ATO processes, address deficiencies proactively, and empower employees to play a direct role in the safety of the NAS by providing valuable safety information. When safety related events occur or are reported the FAA uses lessons learned as an opportunity to train air traffic controllers as to what went wrong, what went right and how can they prevent this occurrence in the future.

## Safety Risk Management

Safety Risk Management is the fourth component of a productive safety culture. Safety Risk Management consists of identifying hazards; analyze, assess, mitigate and accept risk; and implement mitigations and monitor risk.

To proactively reduce the potential for accidents and incidents and to ensure that an acceptable level of safety risk is established and maintained, the ATO uses an integrated safety management approach to expand the perspective of safety analyses. As elements of the Next Generation Air Transportation System (NextGen) are introduced into the NAS, the application of integrated safety management will provide a more thorough approach to performing safety analysis and leveraging existing safety policy and methodologies.

During the risk management process the ATO monitors and manages the top 5 high-priority factors that contribute to the risk in the NAS. The

Top 5 is determined based on data obtained from the Risk Analysis Process, Voluntary Safety Reporting Programs, and other databases used to log and report unsafe occurrences.

Fatigue being a high risk factor for ATC, and aviation overall, the ATO has created the Fatigue Risk Management group. This group provides operational fatigue risk expertise, guidance, and support to the ATO in developing fatigue reduction strategies and policy recommendations to mitigate and manage operational fatigue risks in the NAS. Fatigue unfortunately is part of a controller's job. Air Traffic Controllers on the front lines must work shift work since the NAS operates 24 hours and seven days a week.

## Air Traffic Safety Action (ATSAP)

In cooperation with its employee labor organizations, the Air Traffic Organization (ATO) has established voluntary safety reporting programs for Air Traffic and Technical Operations employees. Air Traffic Safety Action Program (ATSAP) and Technical Operations Safety Action Program (T-SAP) are modeled after the Aviation Safety Action Program (ASAP), created by the airlines. They allow employees to voluntarily identify and report safety and operational concerns as part of the FAA's overall safety goals. The collected information is reviewed and analyzed to facilitate early detection and improved awareness of operational deficiencies and adverse trends.

The primary purpose of ATSAP is to identify safety events and implement skill enhancements and system-wide corrective actions to reduce the opportunity for safety to be compromised. Information obtained from ATSAP will provide stakeholders a mechanism to identify actual and potential risks throughout the NAS. The program fosters a voluntary, cooperative, non-punitive environment for open reporting of safety concerns. ATSAP reports allow all parties to access valuable safety information that may otherwise be unavailable.

Reports submitted through ATSAP are brought to an Event Review Committee which reviews and analyzes the submitted reports, determines whether reports require further investigation, and identifies actual or potential problems from the information contained in the reports and proposed solutions. All Event Review Committee determinations are made by consensus. The Event Review Committee may direct skill enhancement or system corrective action and is responsible for follow-up to determine that the assigned actions are completed in a satisfactory manner. Safety Risk Management may be required for corrective actions.

Compliance with the current versions of the following documents is integral to and supports the successful execution of ATSAP. The highlighted documents can be found on the website https://www.faa.gov/air_traffic/publications/. Documents not highlighted are not available to the public.

- **The ATO SMS Manual**
- The SRMGSA
- Order JO 1030.1, Air Traffic Organization Safety Guidance
- Order 1100.161, Air Traffic Safety Oversight
- **ATO Order JO 3120.4, Air Traffic Technical Training**
- Order 8000.369, Safety Management System
- Order 3000.57, Air Traffic Organization Technical Operations Training and Personnel Certification
- **Order JO 7210.632, Air Traffic Organization Occurrence Reporting**
- **Order JO 7210.633, Air Traffic Quality Assurance Program (QAP)**
- **Order JO 7210.634, Air Traffic Organization (ATO) Quality Control**
- **Order JO 7200.20, Voluntary Safety Reporting Programs (VSRP)**
- Order JO 1030.3, Initial Event Response
- **Order 7050.1, Runway Safety Program**
- **Order JO 7200.21, Partnership for Safety**
- Order JO 1030.7, Air Traffic Organization Fatigue Risk Management
- Order 8040.4, Safety Risk Management Policy

## FAA Safety Top 3 Concerns

The three most crucial safety issues for the Federal Aviation Administration (FAA) were discussed by a panel of experts at the Communications for Safety (CFS) conference in Las Vegas. These safety issues are listed below:

- Runway Incursions
- Soliciting and Issuing Weather
- Issuing Traffic and Safety Alerts

After a discussion of these safety issues we will list best practices for pilots, controllers and airport personnel to increase safety within the NAS.

## Runway Safety

The Federal Aviation Administration's (FAA) top priority is maintaining safety in the National Airspace System (NAS). Safety in the NAS hinges on maintaining integrity, security, and efficiency where multiple safety responsibilities converge, the nation's airports. The goal for runway safety is to improve safety by decreasing the number and severity of runway incursions (RIs) and serious surface incidents.

Runway Incursions are still the FAA's Terminal Tower's largest safety issue. Of all runway incursions (CAT A–D) 52% happened at the middle third of the runway, 90% of (CAT A–B) incursions happen at the middle third.

Runway safety is a significant challenge for everyone in aviation. In the United States, an average of three runway incursions occur daily. Each of these incidents have the potential to cause significant damage to both persons and property. Runway incursions are a serious safety concern and have involved air carrier aircraft, military aircraft, general aviation (GA), and pedestrian vehicles. (Figure 3) Several runway incursions have resulted in collisions and fatalities. Fatalities have occurred at both towered and non-towered airports. A few seconds of inattention can cause a runway incursion.

**There are four categories of runway incursions:**

- **Category A** is a serious incident in which a collision was narrowly avoided
- **Category B** is an incident in which separation decreases and there is a significant potential for collision, which may result in a time critical corrective/evasive response to avoid a collision.
- **Category C** is an incident characterized by ample time and/or distance to avoid a collision.
- **Category D** is an incident that meets the definition of runway incursion such as incorrect presence of a single vehicle/person/aircraft on the protected area of a surface designated for the landing and take-off of aircraft but with no immediate safety consequences.

**Figure 15.3** Runway Incursions are a top FAA Safety Concern that Involves Pilots, Air Traffic Control (ATC), and Ground Operations

A runway incursion is formally defined by the FAA as "any occurrence at an aerodrome involving the incorrect presence of an aircraft, vehicle, or person on the protected area of a surface designated for the landing and takeoff of aircraft." The following are examples of pilot deviations, operational incidents (OI), and vehicle (driver) deviations that may lead to runway incursions.

### Pilot Deviations:

- Crossing a runway hold marking without clearance from ATC
- Taking off without clearance
- Landing without clearance

### Operational Incidents (OI):

- Clearing an aircraft onto a runway while another aircraft is landing on the same runway
- Issuing a takeoff clearance while the runway is occupied by another aircraft or vehicle

### Vehicle (Driver) Deviations:

- Crossing a runway hold marking without ATC clearance

Detailed investigations of runway incursions over the past 10 years have identified three major areas contributing to these events:

- Failure to comply with ATC instructions
- Lack of airport familiarity
- Nonconformance with standard operating procedures

In 2015, the FAA published the 2015–2017 National Runway Safety Plan to address the concerns over runway incursions. This report can be found at https://www.faa.gov/air_traffic/publications/. Since the publication of the 2012 National Runway Safety Plan the aerospace industry has grown more technically complex, undergone a multiplicity of organizational changes, and experienced a rapid surge of multiple types of safety data. To address these challenges, the 2015–2017 National Runway Safety Plan outlines the FAA's strategy to adapt its runway safety efforts through enhanced collection and integrated analysis of data, development of new safety metrics, and leveraged organizational capabilities. The Plan describes the FAA's strategic activities, programs, and objectives associated with achieving the agency's runway safety goals and targets, including the evolution of a corporate approach to managing safety on the nation's runways.

The Plan focuses on the development of the interagency strategic processes to transition from event-based safety to risk-based safety using multiple data sources and stake-holder subject matter experts to assess current risk, predict future risk, and establish relevant metrics that measure the reduction in risk.

In addition to reducing the rate and severity of surface events, another key success metric for the FAA is the measure of how many causal and contributory issues have been identified and corrected.

The objectives of the plan are integration of runway safety efforts consistent with the maturation of the FAA's Safety Management System, establishment of a National Focus Airports Program, further development of runway safety metrics which identify and rate the effectiveness of the agency's runway safety risk assessment efforts, redefinition of FAA organizational responsibility for runway safety, and to further develop internal and external communication and stakeholder engagement strategies to include collaborative training, local leadership and the expanded use of mobile technology and social media.

Several existing safety databases and information collection systems have contributed to the FAA's success in improving safety on the nation's airports. Technology has also played a role in improving runway safety. The technology available is listed below:

- **Runway Status Lights (RWSL)** – system derives traffic information from surface and approach surveillance systems and illuminates red in-pavement airport lights to signal a potentially unsafe situation. Runway Entrance Lights (REL) are deployed at taxiway/runway crossings and illuminate if it is unsafe to enter or cross a runway. Takeoff Hold Lights (THL) are deployed in the runway by the departure hold zone and illuminate red when there is an aircraft in position for departure and the runway is occupied by another aircraft or vehicle and is unsafe for takeoff.
- **Airport Surface Detection Equipment, Model 3 (ASDE-3)/Airport Movement Area Safety System (AMASS)** – is a radar-based system that tracks ground movements and provides an automatic visual and audio alert to controllers when it detects potential collisions on airport runways. ASDE-3 is the radar. AMASS is the software and hardware enhancement to the ASDE-3 radar that provides automated alerts and warnings to controllers.
- **Airport Surface Detection Equipment, Model X (ASDE-X)** – integrates data from a variety of sources, including radars, transponder systems and Automatic Dependent Surveillance – Broadcast (ADS-B) to provide accurate target position and identification information and thus give controllers a more reliable view of airport operations. ASDE-X provides

tower controllers a surface traffic situation display with visual and audible alerting of traffic conflicts and potential collisions.

- **Runway Safety Areas (RSA)** – More than 1,000 runway ends at 500 airports were improved to offer enhanced safety at the nation's airports. The RSA is typically 500-feet wide and extends 1,000-feet beyond each end of the runway. It provides a graded area in the event that an aircraft overruns, undershoots, or veers off the side of the runway.

- **Engineered Material Arresting System (EMAS)** – An EMAS arrestor bed can be installed to help slow or stop an aircraft that overruns the runway, even if less than a standard RSA length is available. EMAS uses crushable material placed at the end of a runway to stop an aircraft that overruns the runway. The tires of the aircraft sink into the lightweight material and the aircraft is decelerated as it rolls through the material.

- **Runway Incursion Mitigation (RIM)** – is a national initiative to identify airport risk factors that might contribute to a runway incursion, and to develop strategies to help airport sponsors mitigate those risks. Risk factors that contribute to runway incursions may include unclear taxiway markings, airport signage, and more complex issues such as the runway or taxiway layout.

- **Electronic Flight Bag (EFB) with Moving Map Displays** – Pilots use Moving Map Displays and Aircraft Own-Ship Position to help them determine where their aircraft is on an airfield, thus reducing the chances of being in the wrong place.

Going forward, the 2015–2017 National Runway Safety Plan, sets the stage for future work to enhance and coordinate runway safety activities. By moving to risk-based decision-making, enabling the safe and efficient integration of NextGen, and demonstrating global leadership in improving air traffic safety and efficiency through data-driven solutions that shape international standards the FAA makes the NAS safer.

Safety is not accomplished by one person or one organization. In order to keep the NAS safe it takes a group effort by pilots, air traffic controllers and airport personnel. Below are listed several best practices for pilots, air traffic controllers and airport personnel. These best practices can be found on the FAA website https://www.faa.gov and the Pilot Handbook.

## Best Practices for AIRFIELD SAFETY – Pilots

1. Encourage use of correct terminology and proper voice cadence.
2. Eliminate distractions in the operational area.
3. Obtain and use airport diagrams. Use the FAA runway safety website to find airport diagrams for all airports.
4. Conduct "Clearing Turns" prior to entering ANY runway.

5. Maintain a sterile cockpit when taxiing.
6. Maintain appropriate Taxi speed.
7. Encourage pilots to have their "eyes out" when taxiing.
8. Encourage pilots to have a "heads up" policy when taxiing.
9. Attend safety seminars and programs on RUNWAY SAFETY.
10. Improve safety by teaching, advocating, stressing and understanding situational awareness.
11. Customize RUNWAY SAFETY presentations for targeted audiences such as pilot organizations, safety seminars, airport authorities, etc.
12. Cite specific airport RUNWAY SAFETY web pages.
13. Distribute RUNWAY SAFETY materials to every aviation entity.
14. Package and distribute runway safety materials to: Flight Schools, Flight Safety International, Maintenance Centers, Aircraft Manufacturers, etc.
15. Realize that every airport is unique and presents its own set of RUNWAY SAFETY challenges.
16. Stay alert; stay alive.
17. Declare war on errors; make it everyone's responsibility.

## Known Best Practices for AIRFIELD SAFETY - Air Traffic Controllers

1. Encourage use of correct terminology and proper voice cadence.
2. Recommend controller usage of the electronic Runway Incursion Device (RID) and the Information Display System (IDS) as an aid to prevent runway incursions. Use the electronic RID with red lamps for runways and amber lights for adjacent areas (mowing, equipment, etc.).
3. Encourage air traffic controllers to tour the airfield, including the runway, taxiway and ramps, during the day, at night and under instrument meteorological conditions (IMC).
4. Encourage locally based organizations to provide familiarization flights for air traffic controllers.
5. Encourage tower cab tours as part of a pilot's training, driver's training and tenant familiarity.
6. Eliminate distractions in the operational area.
7. Air traffic and airport operations should meet following each snow removal day and/or any other unusual event to discuss lessons learned.
8. Develop and publish airport diagrams for ALL towered, commercial and busy general aviation airports.
9. Routinely check airport diagrams for accuracy and update as necessary.
10. Know who has access to the airfield.
11. Update the airport remarks section in the Airport Facility Directory with all applicable data including runway safety information.

12. Determine and publish "line-of-sight" restrictions — can aircraft at opposite ends of the runway see each other?
13. Increase awareness and advertise of local wildlife issues.
14. Determine and publish weather phenomena related visibility issues.
15. Inform AFSS if there is a change in runway status.
16. Encourage pilots to turn lights ON during Landing and Departure.
17. Encourage pilots to have their "eyes out" when taxiing.
18. Encourage pilots to have a "heads up" policy when taxiing.
19. Encourage local flight schools to emphasize runway safety during initial and recurrent training & BFRs.
20. Attend safety seminars and programs on RUNWAY SAFETY.
21. Customize RUNWAY SAFETY presentations for targeted audiences such as pilot organizations, safety seminars, airport authorities, etc.
22. Improve safety by teaching, advocating, stressing and understanding situational awareness.
23. Cite specific airport RUNWAY SAFETY web pages.
24. Use Hot Spot brochures.
25. Distribute RUNWAY SAFETY materials to every aviation entity.
26. Package and distribute runway safety materials to: Flight Schools, Flight Safety International, Maintenance Centers, Aircraft Manufacturers, etc.
27. Realize that every airport is unique and presents its own set of RUN-WAY SAFETY challenges.
28. Stay alert; stay alive.
29. Declare war on errors; make it everyone's responsibility.

## ALL OPTIONS SAFETY TIPS

1. "If you are involved in the problem, you must be involved in the solution" – Christopher Hart, Board Member, NTSB
2. Slow Down to Work Fast..." – Pilot/Controller Panel
3. Standardize your Work" – Pilot/Controller Panel

## Known 'Best Practices' for AIRFIELD SAFETY - Airport Personnel

1. Eliminate distractions in the operational area.
2. Air traffic and airport operations should meet following each snow removal day and/or any other unusual event to discuss lessons learned.
3. Eliminate confusing call signs for vehicles operating in the airport operations area.
4. Maintain a well-defined mowing plan and procedures, including specific area "Designations".
5. Use a patch, or spot system, for mowing and/or farming operations.

6. Use two vehicles for runway inspections to reduce "Time-on-Runway".
7. Use high visibility vehicles to increase conspicuity for pilots, controllers and other drivers operating on the AOA (airport operations area).
8. All vehicle lights (high beams, flashers, beacons, and strobes) should be turned on when crossing or operating on runways, taxiways or the AOA.
9. Vehicle flashers and beacons help ATC, aircrews and other vehicle operators see vehicles in the AOA — especially during periods of reduced visibility and at night.
10. Airport authority should distribute current airport diagrams to all airport users — especially FBO's for transient and student pilots and to other users within 50-100 miles of busy GA airports.
11. Airport authority should coordinate with local fire department, ARFF, and associated training for access to the airfield. Create a "Letter of Agreement" on staging points, alert drills, etc.
12. Re-designate confusing taxiways.
13. Eliminate problem runways.
14. Use current diagrams in all AOA access vehicles.
15. Carry a current airport diagram with all AOA personnel badges.
16. Obtain and use airport diagrams. Use the FAA runway safety website to find airport diagrams for all airports.
17. The airport authority is encouraged to share its driver's training program. (Even FAA employees are required to take training if they are on the airfield.)
18. Utilize CD-based pilot and driver's education training materials and electronic programs.
19. All AOA access authorized personnel, including taxi-qualified mechanics, should complete a driver's training program — to include recurrent training.
20. Require and schedule FAA employee driver's training and recurrent training/testing.
21. Ensure on-airport farming operators are trained and aware of airport operations and its inherent dangers. Ensure farmers know and adhere to agricultural leased boundaries.
22. Encourage inclusion of surface safety training in maintenance school curriculum for taxi and/or tow-qualified mechanics.
23. Offer training and awareness education to local contractors working on the airport, and monitor them.
24. Ensure drivers know where to look for traffic when a pilot isn't talking to the tower or broadcasting on CTAF.
25. AOA access authorized personnel should have an awareness and understanding of the "uniqueness of helicopter operations".
26. Conduct "Clearing Turns" prior to entering ANY runway.

27. Place signs and marking placards in all AOA access vehicles.
28. Know who has access to the airfield.
29. Maximize controlled access to the airfield, including wildlife.
30. Enforce a "No Tailgating" policy to ensure vehicles remain within proximity until gate is closed and secure to prevent unauthorized "Tailgating".
31. Inform the public. Get signs up, "NO TRESPASSING". Enforce "No Trespassing" through ordinance.
32. Keep the runway a runway, no racing.
33. Conduct opposite flow runway inspections. Runway inspections should be conducted toward the flow of aircraft landing and departing as much as possible.
34. Enforce maximum use of existing service roads; stay off of the runway as much as possible.
35. Build and maintain access roads to Navaids from service roads or taxiways, not from runways.
36. Use tunable radios.
37. Enforce a policy of "No Cell Phone" use for personnel while operating on the airfield.
38. Install and/or remove additional signs (including surface painted) and markings to eliminate confusion.
39. Create an airport sign plan and adhere to it.
40. Use lighted runway closure markers to warn pilots of a closed runway.
41. Install signs at the entry point to the AOA and runway safety areas.
42. Prevent potential obstructions.
43. Use standardized "12 inch" and highlighted hold position markings.
44. Maintain runway and taxiway markings.
45. Install elevated runway guard lights (ERGL's) at known Hot Spots and/ or high risk intersections.
46. For new construction, use in-pavement runway guard lights (RGL) at known Hot Spots and/or high risk intersections.
47. Update the airport remarks section in the Airport Facility Directory with all applicable data including runway safety information.
48. Determine and publish "line-of-sight" restrictions — can aircraft at opposite ends of the runway see each other?
49. Increase awareness and advertise of local wildlife issues.
50. Advertise seasonal crops, which might affect line-of-sight for pilots.
51. Issue NOTAMS for snow removal operations and mowing operations.
52. Designate and publish a "Calm Wind" runway at part-time and non-towered airports.
53. Advertise crop dusting operations in the area.
54. Encourage CTAF usage when the airport is "Non-Towered" in the AFD, Hot Spot Brochure, Airport Website, and Posters at ALL on-site facilities.

55. Encourage local flight schools to emphasize runway safety during initial and recurrent training & BFR's.
56. Encourage pilots to have a "heads up" policy when taxiing.
57. Use follow-me vehicles when the ramp is unusually close to a runway and/or for a confusing taxiway route.
58. Attend and conduct safety seminars and programs on RUNWAY SAFETY.
59. Improve safety by teaching, advocating, stressing and understanding situational awareness.
60. Cite specific airport RUNWAY SAFETY web pages.
61. Use Hot Spot brochures.
62. Distribute RUNWAY SAFETY materials to every aviation entity.
63. Package and distribute runway safety materials to: Flight Schools, Flight Safety International, Maintenance Centers, Aircraft Manufacturers, etc.
64. Realize that every airport is unique and presents its own set of RUNWAY SAFETY challenges.
65. Stay alert; stay alive.
66. Declare war on errors; make it everyone's responsibility.
67. Look for runway incursion potential when reviewing airport construction safety plans, especially for haul routes.
68. Always think SAFETY FIRST.

## Soliciting and Issuing Weather

Soliciting and Issuing Weather is the FAA's Enroute/Terminal RADARs largest safety issue.

"…if your soliciting PIREPs just to check the 7110.65 box requirement, you are in the wrong profession, know why you are soliciting and issuing them for the safety of all users in the NAS" – Terry Biggio, VP, Safety and Technical Training FAA.

The FAA, a part of the Department of Transportation (DOT), provides a safe, secure, and efficient airspace system that contributes to national security and the promotion of U.S. aerospace safety. As the leading authority in the international aerospace community, the FAA is responsive to the dynamic nature of user needs, economic conditions, and environmental concerns.

The FAA is responsible for providing a wide range of services including meteorological data to stakeholders of the NAS. The following is a description of those FAA facilities that are involved with aviation weather and pilot services:

The Aviation Weather Center (AWC) in Kansas City, MO issues a suite of aviation weather forecasts in support of the NAS including: Airman's Meteorological Information (AIRMET), significant meteorological information

(SIGMET), Convective SIGMETs, Area Forecasts (FA), Significant Weather Prognostic Charts (low, middle, and high), National Convective Weather Forecast (NCWF), Current Icing Product (CIP), Forecast Icing Product (FIP), Graphical Turbulence Guidance (GTG), and Ceiling and Visibility Analysis (CVA) product. The AWC is a Meteorological Watch Office (MWO) for the International Civil Aviation Organization (ICAO).

Because weather is the most common reason for air traffic delays and re-routings, National Weather Service (NWS) meteorologists support the Air Traffic Control Systems Command Center (ATCSCC) in Washington DC. These meteorologists, called National Aviation Meteorologists (NAM), coordinate NWS operations in support of traffic flow management within the NAS.

An Air Route Traffic Control Center (ARTCC) is a facility established to provide ATC service to aircraft operating within controlled airspace, principally during the enroute phase of flight. Enroute controllers become familiar with pertinent weather information and stay aware of current weather information needed to perform ATC duties. Enroute controllers advise pilots of hazardous weather that may impact operations within 150 nautical miles (NM) of the controller's assigned sector(s), and may solicit PIREPs from pilots.

Airport Traffic Control Tower (ATCT) and Terminal Radar Approach Control (TRACON are terminal facilities that uses air/ground communications, visual signaling, and other devices to provide ATC services to aircraft operating in the vicinity of an airport or on the movement area.

Terminal controllers also must become familiar with pertinent weather information and stay aware of current weather information needed to perform ATC duties. Terminal controllers advise pilots of hazardous weather that may impact operations within 150 NM of the controller's assigned sector or area of jurisdiction and may solicit PIREPs from pilots. ATCTs and TRACONs may opt to broadcast hazardous weather information alerts only when any part of the area described is within 50 NM of the airspace under the ATCT's jurisdiction. The tower controllers are also properly certified and act as official weather observers, as required.

An automated terminal information service (ATIS) is a continuous broadcast of recorded information in selected terminal areas. Its purpose is to improve controller effectiveness and to relieve frequency congestion by automating the repetitive transmission of non-controlled airport/terminal area and meteorological information.

A Flight Service Stations (FSS) provides pilot weather briefings, enroute weather, receive and process IFR and VFR flight plans, solicit and disseminate pilot reports and urgent pilot reports, relay ATC clearances, and issue Notices to Airmen (NOTAM). They also provide assistance to lost aircraft and aircraft in emergency situations, as well as conduct VFR search and rescue services.

The ultimate users of aviation weather services are pilots, aircraft dispatchers, and air traffic management (ATM) and air traffic controllers. Maintenance personnel may use the service to keep informed of weather that could cause possible damage to unprotected aircraft.

Pilots contribute to and use aviation weather services. PIREPs help other pilots, dispatchers, briefers, and forecasters as an observation of current conditions. Pilots should report any observation, good or bad, to assist other pilots with flight planning and preparation. If conditions were forecasted to occur but not encountered, a pilot should also report the observed condition. This will help the NWS verify forecast products and create more accurate products for the aviation community.

The requirements on how to request, receive and issue PIREPs is documented in the Air Traffic Controllers Manual JO 7110.65 and needs to be followed to help keep pilots and passengers safe. PIREPS are the most current and update information a controller can obtain. Even though pilots get forecasts, weather updates and an ATIS, they are not as current as a PIREP.

## Issuing Traffic and Safety Alerts

Issuing Traffic and Safety Alerts is the FAA's Enroute/Terminal RADARs second largest Safety Issue, reports show that Controllers are failing to issue basic traffic calls.

Unless an aircraft is operating within Class A airspace or omission is requested by the pilot, when work load permits, controllers need to issue traffic advisories to all aircraft (IFR or VFR) on the controllers frequency when, in the controllers judgment, their proximity may diminish to less than the applicable separation minima. Where no separation minima applies, such as for VFR aircraft outside of Class B/Class C airspace, or a TRSA, issue traffic advisories to those aircraft on your frequency when in your judgment their proximity warrants it. Provide this service to radar identified aircraft.

When requested by the pilot, issue radar vectors to assist in avoiding the traffic, provided the aircraft to be vectored is within your area of jurisdiction or coordination has been effected with the sector/facility in whose area the aircraft is operating. If unable to provide vector service, inform the pilot.

The issuance of a safety alert is a first priority once the controller observes and recognizes a situation of unsafe aircraft proximity to terrain, obstacles, or other aircraft. Conditions, such as workload, traffic volume, the quality/limitations of the radar system, and the available lead time to react are factors in determining whether it is reasonable for the controller to observe and recognize such situations. While a controller cannot see immediately the development of every situation where a safety alert must be issued, the controller must remain vigilant for such situations and issue a safety alert when the situation is recognized.

Recognition of situations of unsafe proximity may result from Minimum Safe Altitude Warning/Enroute (MSAW/E) or Minimum Safe Altitude Warning (MSAW) for terminal, automatic altitude readouts, Conflict/Mode C Intruder Alert, observations on a Precision Approach Radar (PAR) scope, or pilot reports. If a controller observes an aircraft in a situation of unsafe aircraft proximity to terrain, obstacles, or other aircraft in another controllers airspace it is their job to notify the other controller to make sure the controller is aware of the situation and has issued the proper safety alert.

## Conclusion

As you can see from the discussion in this chapter, safety does not happen with one individual. In order to keep our NAS safe it takes pilots, air traffic controllers, airport personnel, mechanics, weather personnel and technicians, to name a few. Everyone from the personnel on the front lines to upper management must buy in or a safety program will not be successful.

## Chapter Questions

1. What are the four components of the ATO SMS?
2. What are the FAA's three top safety concerns?
3. What is the definition runway incursion?
4. What are four categories of runway incursions?
5. What are the three types of runway incursions?
6. Name three technology devices that have made runways safer.
7. What is a PIREP?

All information in this chapter is consistent with the information and guidance contained in the documents listed below:

- FAA Website
  https://www.faa.gov/air_traffic/by_the_numbers/

- FAA-H-8083-25B, Pilot's Handbook of Aeronautical Knowledge
  https://www.faa.gov/regulations_policies/handbooks_manuals/aviation/
  phak/media/pilot_handbook.pdf

- FAA, ATO SMS Safety Management System Manual July 2017
  https://www.faa.gov/air_traffic/publications/

- FAA National Runway Safety Plan 2015 – 2017
  https://www.faa.gov/airports/runway_safety/publications/

- FAA Fact Sheet – Runway Safety
  https://www.faa.gov/news/fact_sheets/news_story.cfm?newsId=14895

- FAA site Runway Safety, Runway Incursions
  https://www.faa.gov/airports/runway_safety/news/runway_incursions/

- Best Practices for Airfields
  https://www.faa.gov/airports/runway_safety/bestpractices.cfm

- AC 00-45H Aviation Weather Services
  www.faa.gov/documentlibrary/media/advisory_circular/ac_00-45h.pdf

- FAA Air Traffic Control Order JO 7110.65
  https://www.faa.gov/air_traffic/publications/

# Key Terms

Active Failures (Chapter 8)

ADM – Aeronautical Decision Making (Chapter 2)

Aircraft Accident Investigation (Chapter 10)

Anti Drug Program (Chapter 13)

ASAP – Aviation Safety Action Program (Chapter 12)

Bird Strike Avoidance (Chapter 6)

Circadian Rhythm (Chapter 4) Collision Avoidance (Chapter 5)

Cost/Benefit Ratio (Chapter 1)

FAA – Federal Aviation Administration (Chapter 1)

Fatigue (Chapter 2)

FDM - Flight Data Monitoring (Chapter 11)

FOQA – Flight Operational Quality Assurance (Chapter 12)

General Aviation Accident Rates (Chapter 14)

Hazardous Attitudes (Chapter 2) Human Factors (Chapter 5)

Latent Failures (Chapter 8)

LOSA – Line Operations Safety Audit (Chapter 12)

Midair Collisions (Chapter 5)

NTSB – National Transportation Safety Board (Chapter 1)

Proactive Safety (Chapter 9)

Reactive Safety (Chapter 9) Risk Assessment (Chapter 2)

Risk Management (Chapter 2) Safety Culture (Chapter 1)

SMS – Safety Management System (Chapter 11)

# Appendix A:
# Production versus Protection: Fierce Competition and Flawed Culture at Boeing

## The Race for Market Share

John Leahy became chief Airbus salesman in 1994, in the middle of Boeing's reign as the primary single-aisle aircraft manufacturer in the world. Mr. Leahy was relentless as a salesman at Airbus. Once, the chief executive of an airline got sick just as a deal was about to close. Mr. Leahy traveled to the man's house, and the executive signed the papers while wearing his bathrobe. "Boeing thinks we are a flash in the pan, but I don't think there's a reason we can't have 50 percent of the market," he said.[1]

The year 2010 brought some highly-anticipated technological improvements to single-aisle aircraft manufacturing. Airlines around the world loved the single-aisle aircraft for their wide utility and relatively low cost to operate, compared to long-haul aircraft. So much, in fact, that the market forecast for single-aisle aircraft deliveries between 2016–2035 were expected to be 26,860 for a $1.36 trillion value, growing at a compound annual rate of 5%.[2] There was so much competition in this area of aircraft manufacturing, that global conglomerates were entering the market with new aircraft offerings, including the COMAC C919 and the Bombardier C Series. To this point, the two largest aircraft manufacturers, Airbus and Boeing, were looking at the decision to either update their popular airframes, or embark in a brand-new aircraft design. Due to the aging in-service fleets of Boeing and Airbus, airlines who were flying these popular designs were closely looking at upgrades or transitioning their fleet to a new manufacturer.

On a February day at the 2010 Singapore Air Show, Airbus delivered a highly-anticipated announcement for a new engine option that would improve efficiency of their A320 aircraft by 15%. This was a significant improvement to their development timeline, since a newly developed single-aisle aircraft would have only brought 3% efficiency improvements, along with increased costs to develop high-volume manufacturing for carbon fibre structures. Additionally, Airbus was already working out kinks to their two-deck A380 superjumbo jet and its feature flagship, the A350, an all-composite twin-aisle aircraft scheduled for delivery in late 2013. [3] So in order to maintain market share in a highly-competitive market, Airbus took flight with their new Airbus A320neo on December 1, 2010. The race was on.

## Dismissing a Rival

At first, Boeing didn't seem bothered by the new Airbus A320neo. At a business meeting in January 2011, the CEO of Boeing's Commercial Airplanes division, told employees that Airbus would probably go over budget creating a plane that carriers didn't really want. From this statement, the strategy was clear: Boeing had a significant market share and could wait until the end of the decade to produce a new plane from scratch. "I don't think we need to get too spun up over the fact that they're making some sales," he said. [1]

For decades, Boeing had dominated the passenger-jet market with their widely popular 737 and 747 airframes. When Airbus entered the market in 1970, they were a blip on the radar, so to speak. However, things changed in 1999 when JetBlue decided to launch with a fleet composed entirely of Airbus A320s. In the early part of the 2000s, more low-cost carriers around the world placed big orders also. By 2005, Airbus had pulled ahead of Boeing in aircraft sales and widened that lead moving into 2010. According to a managing director of an aviation consulting firm, "Boeing was completely arrogant in dismissing the viability of the A320." [1] And why not? Boeing's culture had been centered around market dominance for decades. Airbus was just a small conglomerate attempting to gain a few percentage points of market share. There's no way they would be able to steal large contracts that Boeing has held for years. Or could they?

## Risk of Losing a Customer

In the spring of 2011, the CEO of American Airlines called Boeing's leader to say they were ready to place an order for hundreds of new fuel-efficient aircraft from Airbus. He went on to say that if Boeing wanted the business, it would need to move aggressively.[1]

Boeing was finally faced with the reality of their decisions. Losing the American Airlines account would be gutting, costing Boeing billions of dollars in lost sales and potentially thousands of jobs. "We won't stand by and let Airbus steal market share," said one executive. However, even as the risk of losing American Airlines loomed, executives at the airline still didn't believe Boeing took the threat of Airbus seriously. [1]

In order to secure the sale, Airbus had a team camped out in a hotel suite in Dallas, near American's headquarters. Senior salesmen traveled to Dallas and dined with chief executives of American Airlines at five-star hotels. On the other side, Boeing visited less frequently. With American still considering which aircraft to purchase, Boeing made a business decision. In order to win over American Airlines, Boeing ditched the idea of developing a new passenger aircraft, which would take a decade to manufacture. Instead, they decided to update it's 737 airframe, promising the aircraft would be complete in six years.[1]

Eventually, American Airlines decided to make deals with both Boeing and Airbus, purchasing 260 Airbus A320neo aircraft and only 100 Boeing 737 MAX aircraft. Roughly three months later, the plan for the 737 MAX was born.[1]

## A New, but Familiar Platform

Now months behind Airbus, Boeing had to play catch-up. Engineers were frantically attempting to design the new aircraft, including compensating for much larger fuel-efficient engines, while trying to accelerate an already tight timeline. Technical drawings and designs were submitted at roughly double the normal pace. A technician who assembles wiring on the MAX said that in the first months of development, rushed engineers were delivering sloppy blueprints to him. He was told that the instructions for wiring would be cleaned up later in the process. When engineers left the MAX project, managers would quickly pull workers from other departments, who were largely unfamiliar with the unique challenges and pressures of the project.[1]

When confronted with the increased time pressure, Boeing released the following in a statement, "The MAX program launched in 2011. It was offered to customers in September 2012. Firm configuration of the airplane was achieved in July 2013. The first completed 737 MAX rolled out of the Renton factory in November 2015." The company added, "A multiyear process could hardly be considered rushed."[1] According to their statement, Boeing's timeline from launch to rollout was a little under four years. Compare this to a similar four-year timeline of Airbus' A320neo, which launched in 2010 and rolled out in 2014. However, one distinct difference is that the Airbus team started the enhancements that were foundational for the A320neo in 2006, which nearly doubles the full timeline of development for the new aircraft to eight years. A much more realistic timeline when considering aircraft development.

Throughout development of the 737 MAX, Boeing maintained its focus on creating a plane that was essentially the same as earlier models of the 737, which was important for getting the aircraft certified quickly. It would also help reduce costs for the airlines, since it would limit the amount of training that pilots would require for qualification in the new aircraft. Engineers were instructed that "any designs could not drive any new training that required a simulator." Pilots were expected to be able to fly the MAX without any additional training.

When upgrading the cabin with digital displays, a lead engineer said his team wanted to resign the layout of information to provide pilots with more data that were easier to read. But since the change may have required new pilot training, they simply recreated the decades-old gauges on the screen instead. "We just went from an analog presentation to a digital presentation," he said. "There was so much opportunity to make big jumps, but the training differences held us back." [1]

## Compensating for New Engines

A pinnacle of both the Airbus A320neo and Boeing 737 MAX was the introduction of new fuel-efficient engines developed by Pratt & Whitney and CFM International. With the fuel-efficiency improvements of the new engines came a cost of increased size and weight. Airbus identified this as a potential issue early in their development by stating, "it's going to be a design change that will ripple through the airplane." [1] But since the new engine was the main selling point of the aircraft, both manufacturers had to manage the design change.

During the design process, Boeing struggled to manage the altered aerodynamics of the new engines, noting that the airplane was more likely to pitch up in some circumstances. To offset this possibility, Boeing added a new software to the MAX, known as the Maneuvering Characteristics Augmentation System (MCAS), which would automatically push the nose down if it sensed the aircraft pitching up at a dangerous angle. The goal was to permit the MCAS system to take over if the aircraft was approaching a stall. Because the system was designed to work in the background, Boeing believed it didn't need to brief the pilots on it, and the FAA agreed. Remember the goal of the new design was that pilots wouldn't be required to train on the new aircraft in simulators. If the FAA or Boeing decided that training on the MCAS system was necessary, they would defeat one of the main production goals of the project.

## Testing and Certifying the New Aircraft

With Boeing's record of developing safe aircraft, the FAA's involvement with the design and development of the 737 MAX was largely a system of checks and balances, rather than one of intense oversight and review. Because of this, the FAA maintained a certain level of trust with the information they received from Boeing and had no reason to investigate the information in great detail. So, when Boeing's chief technical pilot for the 737 MAX lobbied the FAA to remove mention of MCAS from the operating manual and pilot training for the MAX, saying the system would only operate in rare circumstances, the FAA allowed Boeing to do so. But from a series of test flights in full motion simulators, it was apparent that the MCAS was activating in more than just "rare circumstances."[4]

Just weeks prior to FAA certification of the aircraft, Boeing technical pilots noted that the MCAS system was "egregious" and "running rampant" during test flights in the simulator. "Why are we just now hearing about this?" one of the pilots said. "I don't know, the test pilots have kept us out of the loop," said the other."[4] Through simulator testing by the technical and test pilots, a unique design element of the new system was found. The MCAS system relied on only one angle of attack sensor, whereas the aircraft had multiple sensors installed. Lack of oversight of this crucial design flaw allowed a single faulty angle of attack sensor to activate the MCAS system and command nose-down pitch, without any prior indication to the pilot.

Among the communication exchanges during the testing process was a concern by Boeing's chief technical pilot that he was "unknowingly lying to the FAA" about the safety of the MCAS system. In separate emails he sent to an FAA official, he said he was "Jedi-mind tricking" regulators outside of the United States into accepting Boeing's suggested training for the MAX, without the requirement for training on MCAS.[4]

Despite the internal struggle during the testing phase, the 737 MAX gained certification from the FAA in March 2017, followed by other regulators worldwide later that month. The first 737 MAX delivery was made in May 2017 and by October 2018, 230 of the aircraft had been delivered to customers around the world.

## Two Disasters Unveil a Flaw

### Lion Air Flight 610

On October 29, 2018, Lion Air flight 610 took off at 6:20am from Jakarta, Indonesia. Despite the pilots observing incorrect speed and altitude readings on a previous trip, the aircraft was kept in service until it could reach a maintenance facility for repairs. As soon as the aircraft took off, the pilots received a stall warning indication. Due to the discrepancies noted on the previous flight, the pilots could not ascertain their speed or altitude, and subsequently told air traffic controllers that they were "experiencing a flight control problem." The nose of the aircraft seemed like it was being forced downward. Twelve minutes into the flight, the aircraft crashed into the Java Sea, killing all 189 passengers on board.[5]

Over the next few days, the investigation began to focus on the MCAS technology and the pilots' reaction to the system. Within a week of the accident, the FAA and Boeing issued an Airworthiness Directive warning about possible trim stabilizer control issues due to faulty angle-of-attack indicators. Although Boeing didn't call MCAS out by name, they argued that the Flight Crew Operations Manual described the function, and how to override it. In the meantime, airlines around the world continued to fly the MAX and Boeing continued to build and deliver more of the aircraft.[5]

## Ethiopian Airlines Flight 302

On March 10, 2019, only five months after the Lion Air accident, Ethiopian Airlines Flight 302 took off at 8:38 am from Addis Ababa, Ethiopia. Almost immediately after takeoff, the pilots reported a flight-control issue. One minute later, MCAS activated, pitching the nose downward. The pilots struggled to control the aircraft, but MCAS activated again. Per the protocol issued after the Lion Air accident, the pilots disabled the electrical trim tab system, but were unable to manually turn the trim wheel, partly because the crew inadvertently left the engines at full thrust after takeoff. Six minutes later, the aircraft crashed into the ground at 700 miles per hour. All 157 people on board perished.[5]

Regulatory authorities again cited MCAS as a contributing cause of the accident, combined with the fact that the pilots could not adjust the stabilizer trim by hand. While there was an electric system to help turn the trim wheel, the system was disabled by the same switch that disabled MCAS.[5]

## Worldwide Grounding

After the accident, Ethiopian Airlines immediately grounded the rest of its 737 MAX aircraft, followed by airlines around the world within the next two days. Among the worldwide regulators, the FAA was the lone holdout, initially reaffirming that the aircraft was airworthy. Before the FAA issued its order to ground the aircraft, President Donald Trump ordered the grounding of all Boeing 737 MAX aircraft on March 13, 2019, three days after the Ethiopian Airlines accident.

## Case Questions

1. How did Boeing balance the production vs. protection dilemma?
2. What internal and external factors contributed to the decision-making process at Boeing?
3. Address the culture at Boeing during the time of production of the 737 MAX.
4. What hazardous attitudes were demonstrated by the Technical Pilots and Engineers of the 737 MAX during production, testing, and certification of the aircraft?
5. Address the chain of events at Boeing during the production and certification of the 737 MAX that constituted the "Swiss Cheese Model" of accident causation for both the Lion Air and Ethiopian Airlines accidents.
6. Finally, if you were an FAA or NTSB investigator, what practical recommendations or regulatory changes would need to happen before allowing the 737 MAX to return to service?

CPSIA information can be obtained
at www.ICGtesting.com ·
Printed in the USA
LVHW021643100521
687008LV00001B/1